REFLECTI

George Hows

This Coat of Arms was granted to the name Hows and its
derivations in medieval times.

PUBLISHED BY

RAWSON PUBLICATIONS

i

REFLECTIONS

ISBN 0-9548469-0-7

First Published in 2004 by
Rawson Publications

Printed by Great Britain Warners Group Publications plc
West Street, Bourne, Lincolnshire PE10 9PH.

REFLECTIONS

Best Wishes
George Hawe.

DEDICATION

For my children and grandchildren that they know something of their dad's and grandad's history

G H June 2004

CONTENTS

ACKNOWLEDGEMENTS

In particular I want to thank Pat Clark who inspired me to write this story, has encouraged me throughout and become my editor.

Secondly to Mike Rawson without whom the work would never have progressed. He taught me the initial use of a computer and living nearby has monitored the work almost daily.

Cyril Hack for all his help and encouragement during the writing of this book without which it would never have reached completion.

My family and friends who provided material and urged me to continue.

Harry Turner a fellow sufferer in children's homes days and a lifelong friend has contributed a great much of the detail in this narrative.

The East Anglian Daily Times for the article entitled Mobile Viners are tried in Australia.

The Daily Telegraph for the 'Ohio' arrival in Grand Harbour photograph.

The Leicester Mercury for the article on the 'Siege Bell' cast in Loughborough.

FOREWORD

A man who sprung from inauspicious background was bounced from one workhouse to another until at sixteen years of age he faced the realistic world with neither relative nor friend except his God.

Apprenticed in the agricultural machinery trade and determined to prove his worth, he proceeded through every facet to become Managing Director and ultimately Chairman of the Company.

He served in the RAF for six years both in the Siege of Malta and in the Italian Campaign servicing Spitfire aircraft.

Raising a family of three and widowed sixteen years ago, he continues to support his church and to take a major role in the Masonic world together with his charitable work with the Melton Volunteer Bureau.

Sharing George's workhouse experiences I have followed his path through life and am privileged to be his friend and appreciative of all his kindness, help and advice throughout all those years.

Harry Turner 2004

Chapter One
Early days

A Sagittarian, a traveller,
to be born so true,
from Alford to South Africa,
from Partney to Peru.

I was born on the 29th November 1921. My mother was a small, petite lady only five feet tall and, although aged only forty her hair had already shown signs of grey with wrinkles on the face. Perhaps it was the result of the worries of raising a family on such a meagre income in the bleak climate of Lincolnshire. She had come up from Ipswich in Suffolk where the climate was much kinder and the

George aged four with Mum

living conditions were very much better. All the houses there had running water and modern sanitation and most were favoured with mains gas and electricity.

Mum's maiden name was Ethel Grayston. She was the eldest of four daughters born to the Grayston family who were engaged in the printing business. Grandpa Grayston was a lovely old gentleman who did not enjoy the best of health but never grumbled and was so very kind to his grandchildren. His hobbies were carpentry and photography and he was very proficient at both.

Of the other three daughters, Mabel remained single and stayed at home, Ivy married a pharmacist, Kenneth Southgate, and they had an only child, a son named Ross. Nellie married a manager in a gentleman's outfitters shop and produced a daughter named Joyce and a son, Peter. This meant that Olive my sister and I had several aunts and uncles on the Grayston side of the family.

My father's name was Walter Househam. He was the eldest of the four sons of George Henry Househam: a tenant farmer at The Elms, Sloothby, near Alford. Lincolnshire. A daughter named Lilian completed the Househam family.

Grandpa and Grandma Grayston with their four daughters. Mum is featured here in the centre of the back row

We lived in a small country cottage at Halton Holgate, near Spilsby. Father worked on a local farm and the house known as a tied cottage went with the job. Accommodation consisted of two rooms up and two down and facilities could only be described as primitive. There was no mains gas, water or electricity and toilet arrangements were basic - a 'privy' at the bottom of the garden.

In 1927 the country, devastated by the General Strike, drove farming into deep depression. Many farmers faced bankruptcy while others cut labour in the struggle to survive. In some parts of the country, landowners waived rents due from tenants rather than allowing their farms to become derelict, while others charged minimum rates, as low as sixpence an acre.

Father was thrown out of work and, living in a tied cottage, was forced to move. He now had no work or home for his family of four. Fortunately his father, farming less than twenty miles away and suffering like others, offered Walter temporary employment but no accommodation. Father searched for a house in the village. He was lucky. A local builder, George Rooke, (whose wife was the village school teacher) had recently purchased a property named High Acre, and converted it into two separate houses. One was to be his own home and the other one was to be let. Father secured the tenancy at a weekly rent of six shillings. Thus our move to Sloothby was planned.

Before we left Halton Holgate, Grandpa Grayston had made for Olive and me a 'super' wooden handcart.

Large enough to carry two children it was complete with big wooden wheels, a long handle and a cushioned seat. Painted dark green with red wheels and a red stripe round the body, it was indeed our pride and joy as it was so superior to those made from an old pram chassis and a couple of boards tied on with string. We flaunted our new possession around the village at every opportunity, taking it in turns as to who should ride. I did not mind pulling as I enjoyed pretending I was a horse. Olive soon tired and lost interest when it was her turn to pull but this was no problem. Other boys soon joined me and made full use of what came to be known as the Rolls Royce cart.

George and Olive with Grandpa's hand cart

'Travel in comfort' was the slogan of this new syndicate. Many other utilitarian uses were to be made of this vehicle in years to come. Although I was under five years old at the time I could see potential in the future. Might I even charge a penny a ride for children not members of the 'syndicate'?

One day when Olive and I returned from school, we found the house in turmoil. In the kitchen Mum was on her knees wrapping crockery in newspaper and carefully packing it in a tinfoil lined tea chest. With her greying hair and saddened face, I perceived a tear roll down her left cheek so I could see that she was upset.

"We're moving tomorrow," Mum said.

I was puzzled and confused by the news. Aged only four and a half I was too young to realise that moving home was one of life's most stressful experiences.

On going into the living room I found it in chaos. The walls were bare. Grandpa Grayston's photograph and Mum's favourite picture of two cats had gone. I rushed upstairs to change into my play clothes and found Dad taking our beds to pieces.

"You'll have to sleep on the floor tonight," he said and carried on working.

I went downstairs and, instead of going out to play, asked Mum if I could do anything to help.

"You'd better get your things together and put them in the half-filled tea chest and tell Olive to do the same."

The few children's books and toys that we owned were duly packed and after our tea we went upstairs to bed. My mattress was on the floor and as I was so tired I slept well.

Next morning Mum woke us at 6 am., and I realised it was the day of the move. I was excited. I drew back the curtains and, in the light of the dawn, observed horses and wagon parked outside. Apparently Father had left home early, cycled to Sloothby and borrowed them from his father's farm.

I washed briefly and, giving my face a quick rub, grabbed the only towel in sight and went through the motions of drying. Excitement was getting the better of me and I needed to get out and see the horses. On dashing downstairs I saw a plateful of sandwiches on the kitchen table that Mum must have prepared the night before. Grabbing a couple, I hastened outside. Dad, helped by a neighbour, was busy loading our possessions on to the wagon.

"Can I help, Dad?"I asked, only to get the brusque reply,

"No, you would only get in the way and be a nuisance. Go upstairs and watch from there."

This I did and observed progress from the bedroom window. It was to be a memory that would last forever and to this day, I never understood why we had to move. By eight thirty all was loaded and a seating space was left at the back for Mum and Olive. Space at the front next to Dad had been made for me.

Scrambling aboard I grazed my knee but as it didn't bleed, I ignored the pain in the excitement. Making myself comfortable I anxiously waited for Mum and Olive to get aboard. At long last they emerged from the house, Mum shrouded in a massive shawl and Olive in a long overcoat at least four sizes too big for her. In my haste I had left my coat in the house but Mum, seeing what had happened, brought it out to me. "Put it on at once" she shouted,

"you'll freeze up there."

I thanked her and obeyed immediately. Mum and Olive struggled aboard making themselves comfortable using an eiderdown and a couple of cushions. After what seemed ages Dad appeared, locked the front door and left the key with the neighbour. He then checked that the load, for which he alone was responsible was safe and secure before taking his place in the driver's seat.

He gathered the reins calling the horses, Prince and Smart by name, and with a peculiar click, click made from the corner of his mouth, urged them into action. They strained forward, the wheels began to turn and we were on our way. We progressed somewhat leisurely and I was fascinated with the clip-clop of the horse's hoofs and the swishing of their tails.

Soon I felt the urge to hold the reins myself.
"Dad, can I have a drive?" I asked.
He seemed to ignore me. I raised my voice just a little and repeated,
"Can I have a go Dad?"
"Alright I'll show you what to do."
"Hold the reins in each hand like this, but not too tight," he replied and demonstrated.
I took the reins and strictly followed his instructions. I now felt myself to be in complete control and was thrilled. I had observed how Dad lightly pulled either rein according to the bend in the road so I followed his example and Prince and Smart responded.

On and on we travelled through Spilsby, Skendleby, Candlesby and Gunby. I could read the village signs as we passed as the pace seemed so slow and I was anxious to reach our destination.
"Dad," I called, "I'm getting tired, and my arms are aching."
"Alright! I'll take over, give me the reins," he answered.
I did and was glad to have a rest.
By mid-day we had completed only eight miles but Dad

confirmed that two miles an hour was all that could be expected of two horses pulling such a heavy load. We had another nine miles to go. We turned left at Gunby Corner on to a much narrower road, which wound its way through Welton-le-Marsh, Habertoft and Hasthorpe. There had been no time to stop and eat, only a five-minute break for the horses. Each had been given a nosebag of oats, and a pail of water obtained from a nearby pump. We had eaten sandwiches washed down with cold tea whilst travelling. I fell asleep and as there was no sound from the back I thought Mum and Olive had done likewise. At about 4 pm., we reached Sloothby and so ended a very tiring journey.

Dad had arranged for his younger brother, George, who lived with his father at The Elms, to help us unload. The men dealt with the furniture while Olive and I carried the smaller bits and pieces into the house. I was more interested in exploring the layout of the rooms.

Downstairs there was a sizeable kitchen with an old fashioned sink. This led to the dining room and beyond a front room to be called the lounge. A wide staircase led upstairs to one large bedroom and two smaller ones. This meant that Olive and I could now have separate rooms, a much better arrangement than hitherto.

By the time that the wagon was empty, darkness had fallen. Dad proceeded to assemble the beds and Olive helped Mum to make them up. All this was done by candlelight until later that evening an oil lamp was pressed into service.

Fortunately Grandma Househam at The Elms, who realised that we could not have had a proper meal that day, prepared a hot meal for us all. On entering the farmhouse kitchen we beheld a huge kitchen table that had seating for at least twenty people, filled with various tureens of vegetables and a massive beef stew and dumplings. All were served, some with a second helping until nothing was left. The dishes were cleared away and Grandma brought in a huge jam roly-poly pudding together with a big jug of custard. I was not the only one who had been desperately hungry and this was my favourite meal, so I enjoyed every mouthful.

After returning to our house and being very tired, I fell into bed and must have been asleep within seconds. Next morning I was still asleep at eight thirty when Mum called me. Olive had already breakfasted and was ready for school. I quickly washed, dressed, grabbed a bowl of porridge and joined her. Luckily, school was only two hundred yards away so we were able to reach the playground as the bell for assembly was ringing. Joining the appropriate column - boys to the left, girls to the right - we marched into the school building, which consisted of one large room to which had been added a smaller one. This was used as the chancel when the building doubled up as a church on Sundays.

There appeared to be but one teacher, Mrs. Rooke, the wife of George Rooke, the builder. Later it transpired there was an assistant, Miss Betty Huxstep. A total of twenty pupils attended, aged from four to fourteen. The school was set back from the road with the playground in front which was equally divided so that boys and girls each had separate play areas.

On returning from school I noticed, on the front of our house the letters A W   A. These were in yellow brick and distinguished from the red, normally used in house building. A window occupied the gap between the third and fourth letters. I have since learned that High Acre was originally built in 1878 as a workhouse by the Alford Workhouse Association. Was the missing letter an 'H?' If so, perhaps it was to perpetuate the names of the founders.

High Acre - Sloothby

Chapter Two
Life at Sloothby

In the following days and weeks I explored the village so as to familiarise myself with the locality and its people. Sloothby is a small Lincolnshire village situated between Hogsthorpe to the east and Willoughby to the west, each being two miles away. From Hogsthorpe it is just five miles to Skegness and from Willoughby a further three miles takes you to the market town of Alford.

This is predominately an agricultural area comprising small farms of fifty to a hundred acres. In addition there are a few smallholdings, the remainder being marshland. The soil is heavy clay and difficult to cultivate. Consequently much was down to permanent pasture and grazed in the summer months by fattening cattle. Sheep are reared for meat and wool and horses for the only source of power available. Pigs and poultry are raised in and around the farmyard. One dairy farm produced sufficient milk for the village and surrounding hamlets.

The butcher, Mr. Carter, came from Hogsthorpe and the baker, Mr. Baker, from Candlesby. Other requirements, not available in the village shop, called for a visit to Alford that had a market day on Tuesdays. Travel was normally by the weekly bus but the farmers usually went by horse and trap. This may be the one with large wheels and known as a high trap or one with much smaller wheels and designated as a tub trap. There wasn't even one car in the village at the time.

Sloothby boasted a general store from which most items of food and drink excluding alcohol, and children's clothes were available. Paraffin for lamps, candles and carbide for cycles were also carried in stock. Accumulators for the wireless could be taken there and re-charged for the princely sum of sixpence. This emporium also doubled up as the local Post Office.

A blacksmith attended the local 'smithy' two days a week providing a farrier service as well as repairs to farm implements. On occasions he could be persuaded to make an iron hoop for an enthusiastic boy as a reward for blowing the forge bellows.

A joiner / wheelwright / undertaker, Mr. Coupland, served most demands which made use of wood. He was a friendly character and, if an inquisitive boy should appear and there was a coffin being made, it was not unusual for him to be provided with a sheet of sandpaper and asked to smooth the lid. I know that's true as I was once caught myself. Needless to say 1 never went in to see him again.

There was also a herbalist in the village by the name of Mrs. Booth. She dispensed a wide variety of pills and potions and claimed to cure many ailments.

The nearest doctor was at Willoughby and only attended serious cases. If you wished to consult him with minor complaints it meant a two-mile walk or a ride should you be so fortunate as to own a bike.

The village pub was called The Fox's Head with Mrs.

Francis, a war widow, the licensee. It was owned by Bateman's of Wainfleet, near Skegness, famous for their Good Honest Ales.

Sloothby's population in those days was approximately eighty including twenty-two children of school age. There was a Chapel with seating for eighty, and the school, which doubled as a Church of England church

Sloothby School - Mission Church

that had been known to seat as many as seventy on special occasions. I remember one event when the Alford Salvation Army Band gave a concert in the Methodist Chapel. They gave a rousing programme of religious songs and marches, which opened with "Trust and Obey". This was the first time I had heard this tune, and it is still one of my favourites. Once a year a Mission Team would arrive, erect a large marquee in Simpson's field opposite the blacksmith's shop and, for a week, conduct Mission services every evening.

Other major functions during the year included the Chapel Anniversary usually in June with presentations from the Sunday school children. In October there would be two harvest festivals, one for the Anglicans and one for the Methodists, both being considered Red Letter days in the village.

In July a Garden Fete took place on Alfie Francis's meadow near the chapel. Early in November, the children sang Guy Fawkes songs around the village to raise funds for their fireworks. At Christmas time a mixed choir of adults and children toured the village singing carols and collecting money for charity.

With these exceptions Sloothby was a quiet and peaceful village, scarcely known to the outside world. Everyone had to make their own entertainment. Boys played 'whip and top', bowled hoops, played 'conkers' or flicked 'fag cards'. The girls skipped, played 'hopscotch' or hockey. The grown-ups sang songs and hymns round a piano on Sunday evenings, while others who owned a wireless would spend their leisure time 'listening in' probably to the service broadcast from St Martin's in the Field where the famous Canon Dick Shepherd was the preacher.

Some ladies would knit or crochet while others practised the art of lace making. Life in Sloothby was somewhat primitive in those days. Water was pumped from a well shared by several households. Sanitation was in the form of a 'privy' (a very basic form of a toilet as it would be known today) at the bottom of the garden. Candles and oil lamps were the only source of lighting, while coal and coke on open fires was the only means of heating water for washing and cooking. A separate fireplace in each room kept the house at a reasonable temperature in winter, all these being extremely labour intensive.

Winters were often long, with severe weather reducing normal life to a standstill. Heavy falls of snow blocked all means of access and 'digging out' became the number one priority. Life on the farms presented the most problems, as livestock had to be fed and the cows milked every day of the year irrespective of the weather.

The farmer and his stockmen started work at 5 am, probably having to clear a path from the house to cowshed, stables and barn. Any frozen pipes then thawed out. Horses would then be foddered and groomed. Milking the cows came next and then mixing food for the pigs and feeding them.

Breakfast was at 7 am, then back to work. Corn was scattered in the farmyard for the free-range chickens, ducks and turkeys (if any) in their respective areas. Lastly the sheep. Turnips had to be sliced and taken down to the field where the sheep grazed. No wonder farmers dreaded the bitter winter weather.

For the children, life took on a new dimension. Snow provided the fun of making snowmen and battles with snowballs. Any nearby snow-covered slopes were used for tobogganing and any flooded areas frozen hard, attracted skaters by the score. But schooling had to continue, at least for those who were able to trudge through the snow, but those living in outlying areas stayed at home severely reducing the numbers in class. It could be days or even weeks before all resumed their schooling.

Much of my time throughout the year was spent with Grandma at the Elms Farm nearby and I enjoyed helping in many of the tasks that I was given. One day after school I went to see Grandma.
She enquired, "Do you want a job, George?"
"Yes, if you like, what is it?" I replied.
"You can collect the eggs," she answered.
"There are the two hen houses and you know where the nests are."
So off I went. I found about a score and returned to Grandma.
"How many have you got?" she asked.
"About twenty," I replied.
"There should be more than that," she retorted.
"There are several nests around the farm. Empty the basket and go and have another look around."
So off I went again and, to my surprise, found nests in the most unusual places. One was in the remains of an old haystack, one in the barn among the sheaves of oats and another among a bunch of nettles by the pond. By now I had collected twenty-nine eggs and rushed back to tell Grandma.
'Look what I have got, Grandma. All these."
"Thank you very much, George. I'll have you doing that job again now that you know where all the nests are."
And so I became the part time egg collector.
Another day Grandma gave me the job of turning the

milk separator.

"You can turn the separator handle today, turn it faster and faster until the bell rings and keep it at that speed," she instructed.

Grandma poured the milk into the vessel at the top of the machine and, after passing through, the cream came out of one spout and the separated milk from another. Grandma then said;

"Now you know what to do, I'll have you doing that again and, if you come on Saturday, I'll show you how to make butter".

By this time Grandad could see me as a potential helper in the farmyard. Consequently, one day in the school holidays, when I was talking to Grandma, he said to me;

"Now young George, I've got a job for you. I need some turnips slicing for the sheep, come with me!"

I followed him to a huge heap of turnips in the yard with another machine standing by.

"All you have to do is to turn the handle and I'll feed the turnips in the top," he told me.

A straw skip was placed below the machine and soon filled with turnips now cut in equal slices.

"That will do fine, young man. You can come and do that again," said Grandad.

So life went on. If Grandma didn't find me a job, Grandad did. I always found a job at The Elms and I learned a lot in the process and what is more, I enjoyed every minute of it.

Being with the farm animals and poultry both young and old taught me so much. The various life cycles of each were different and fascinated me. Many ended their days on the breakfast or dinner table. I learned the story of the pig that could appear as bacon, pork, ham or sausage. Cattle provided us with veal, beef, milk, cream and butter. I now realise how fortunate I was to have all these experiences at such a tender age.

However, life is not all learning or all work. The pursuit of sports and games had its share of my time. Bowling a hoop was one game popular before the advent of the motorcar. It was played on the roads with a caned hoop, scrounged from an empty orange barrel at Alford fruit market and a stick cut from a hedge.

Spinning tops was another favourite. A top bought for two pence at the village shop and a whip made from a stick, cut from a hedgerow, fitted with a leather bootlace was all that was required to play 'whip and top'. It was not an easy game to play and could require very many attempts, winding the lace round the top and pulling it quickly before the top began to spin. Once spinning further strikes with the lace would, or should, keep it spinning. It was, however, a skill that took some time to acquire.

My days at Sloothby School were numbered as, when I reached the age of ten, it closed. Teachers and pupils were transferred to Willoughby, just two miles away. A school bus bound for Alford collected us and dropped us at the new school gate. Nearly one hundred pupils attended this school making the classes somewhat larger than at Sloothby but the curriculum was similar. Our sojourn was to last but twelve months for, at the age of eleven, all were moved to a new area school at

Alford. The same bus now took us to the school gate.

This was an entirely new experience – modern building, new teachers and a headmaster who, at first sight appeared to have a Dickensian approach to discipline. The curriculum was greatly extended and included science, woodwork, metal work and gardening for the boys and housewifery, needlework and cookery for the girls.

Although very strict the headmaster, Mr. Arthur Moore, took great interest in his pupils and had tremendous pride in his school. Nothing but the best would satisfy him. He was determined to have it recognised as the best school in the East Midlands.

At assembly one morning he announced that he had entered the school in a Choirs Festival and what was more, Alford School would be the winners. At that point in time no such choir even existed. Being an accomplished musician himself he trained a choir of forty voices. First, he auditioned eighty boys and girls and commenced practices after school hours so as not to impinge on the singer's education. Progressively he de-selected those not up to the required standard until the required forty remained, of which I was one. We sang "The Ash Grove", "Who is Sylvia" and Blake's "Jerusalem".

The Festival was held at Cleethorpes and, to our surprise, Alford School were the winners. Imagine the pride of our Headmaster as he mounted the rostrum to receive the silver trophy. His dedication to the choir was well rewarded and acclaimed by the whole school on our return. This was typical of Mr. Moore's attitude to the teaching profession, tough he might be but fair and he demanded near perfection. No wonder he earned for the school a reputation second to none.

I enjoyed my days there until a bombshell fell into my lap. It happened like this. One day a boy, unknown to me, approached and asked if, on the following day, he could take me to see his granny. As she lived in Alford we could go in the lunch hour. I agreed and the next day we crossed the town to West Street. There, at Number 68 sat an old lady in the open doorway.

"Hello!" she said. "Is your name George?"

"Yes," I replied.

This petite lady with greying hair and beady eyes seemed to look right through me. She was searching for something and I knew not what. Then, after what seemed an eternity of silence she asked:

"When is your birthday?"

"It's in November." I replied

"It's the 29th, isn't it?" she asked.

"Yes." I admitted.

But how did she know? I wondered.

Yet again her beady eyes searched about my face.

"Yes," she muttered, "I'm certain I remember the face - those deep set eyes and prominent forehead features never change. What did you say your name was?"

"George Househam," I replied

"No, no, your name is Hows not Househam.

Imagine, eleven years old and hearing a statement like that – I was shocked.What was the old lady going to tell me?

Chapter Three
Revelation and Tragedy

She continued; "About eleven years ago a young woman carrying a baby in a shawl knocked on my open door. When I saw her I could sense that she was distressed so I invited her in and gave her a cup of tea. Suddenly she burst into tears and then, hesitantly, told me of her dilemma. She was a barmaid in a Nottingham 'pub' and had met a serving soldier stationed locally. A friendship developed and they became lovers resulting in her pregnancy. Convinced that they were to be married she continued with the pregnancy and a little boy was born and, as so often is the case, the father left her in the lurch and sought a posting overseas."

"Unable to trace him she was left on her own with the baby. Her job at the 'pub' was terminated and she was left homeless and with no means to support either herself or the child. She pleaded with me for help and offered to pay me two shillings a week if I would take the child and care for him. I was a widow and took pity on her so agreed. I just did not have the heart to ignore her plea."

"For several weeks an envelope arrived in the post containing a two shilling Postal Order but no note to indicate her whereabouts. Then the payments ceased and, being myself of limited means, I had no option but to report the matter to the local police. They informed me there was no choice but to place the child, now about six months old, in the care of the Lindsey County Council, which meant the Workhouse. That is all I know. And it wasn't until my grandson told me of your arrival at school that I guessed it was you."

I was devastated and shocked. Was it really true? Was not Mum my real Mum? Or Dad my real Dad? I was broken-hearted. But it was already 1.20pm., and we had to get back to school. All afternoon I was confused and unable to concentrate on writing an essay during the English class. I just wanted to get home and ask Mum if all this was true. I had only ever known Mum and Dad who had loved and nurtured me all my life. The same applied to Olive who I believed to be my sister. There must be some explanation.

The school bus seemed to take hours before we reached Sloothby and I was able to get home. I tore into the house crying:
"Mum, mum, where are you?"
She was upstairs and called out:
"I'm up here. What do you want?"
"Come down, mum, I must talk to you," I replied.
She came down immediately realising that something must be wrong. I burst into tears and she exclaimed.
"Whatever is wrong with you?"
I asked. "Aren't you my real mum?"
"I went to see an old lady in Alford today and she told me everything. She said that my real mum abandoned me when I was very young and I was taken to the Workhouse at Spilsby. "Is that true, mum?"
Mum too was upset and started to cry.
"We had always intended to tell you everything but not until you were older and able to understand. But now I must tell you. You see, George, your Dad and I were

unable to have a family of our own and this was a great disappointment to us both. After making enquiries of our doctor he suggested that we should contact the Lindsey County Council who might be able to help. This we did and were informed that young babies were sometimes committed to their care. A nursery existed at Spilsby where babies were accommodated up to the age of two years. They were then boarded out to childless families in the area who became foster parents. At the age of fourteen the foster parents were given the option to adopt any children still in their care. This arrangement was subject to termination at any time the children didn't settle down or if the foster parents were unable to cope. Remuneration at the rate of six shillings a week was paid to the family for participation in this scheme."

"Olive came to us when she was two years old. She soon settled in and your Dad and I were delighted to have a child in the home. A year later at the same age you joined the family too. It was our intention that you would be adopted but, at this stage we could not tell you. Now you know the whole story. We love both Olive and you and have no intention of breaking up the family."

This explains why I believed Olive to be my sister earlier. Obviously the whole situation hurt me a great deal but there was nothing either Olive or I could do to change it. But there were more serious implications. There was no mention of my birth parents ever getting together or getting married. Therefore I was illegitimate, a bastard, and would carry that stigma all my life. In those days it could be a serious handicap when applying for work or endeavouring to make headway in society. Today, thank God, no such prejudice seems to exist.

Times were very hard. Due to the state of the national economy farm prices fell even further and Grandad Househam had no option but to reduce his labour costs. Dad's temporary job had to go. He was now on the dole. He sought temporary work wherever it could be found. Cycling around the area he found only seasonal jobs. Sheep shearing gave him a couple of days, horse clipping a few hours here and there. On one occasion he 'travelled a stallion' for a few weeks. Then there was the odd day's work pig-killing. But there were so many days when, in spite of travelling miles around, he came home with empty pockets.

He then tried something entirely different. He took an agency for Fulljoy sponge cakes. Fitting carrier boxes front and rear to contain the cakes he cycled for miles endeavouring to sell them but met with very little success. He had to look even further for work that would provide him with a regular income.

He now considered selling daily papers. In his travels around, so often folk in the out-lying areas had commented on their difficulty in getting a paper or a Radio Times. Perhaps there was an opportunity there. Dad decided to have a go. He bought twenty papers - Daily Express, Daily Mail and Daily Mirror – and went up to Hasthorpe, Habertoft and Welton and sold the lot. In fact, he could have sold more. On returning home he contacted a wholesaler and ordered fifty papers and again sold out. He was beginning to build up a regular business. His round extended to twenty-five miles a day and serving nearly a hundred families. He was now selling periodicals, which added to his earnings without extending his mileage.

Many customers now indicated their need for a Sunday paper, another opportunity not to be missed, though it did mean working a seven-day week. This was real progress. The cycle was replaced with a motorbike and sidecar and it was at this stage that I became interested.
I asked. "Could I go for a ride in the sidecar on a Saturday, Dad?"
Dad replied "Of course you can."

He had every reason to agree. This would not be just a joyride for George; he could be made good use of, helping with the deliveries. Riding in the sidecar really was an exciting experience. Bumpy it might be but when Dad was cornering it seemed as though I was about to be airborne. Dad soon had me delivering the papers and, although I was not yet even eleven, to collect the payments due.

The business continued to progress as more villages were added to the round. It was then realised there was a demand for 'The Green Un', or as it should have been named "The Football Echo" which was printed on green paper at Grimsby. It was a sports paper that gave the latest football results. This meant collecting them at Ulceby Cross, the other side of Alford, and stopping only at fish shops and pubs to leave the required numbers.

By now, the motor bike was superseded by a Ford saloon car with petrol costing one shilling (5p) per gallon (4.5litres). In the summer, yet another opportnity presented itself. Scores of coaches and charabancs poured into Skegness on Sundays bringing hundreds of visitors from the Midlands, so we made a bee-line for the Coach Car Park which would be the venue of our next 'point of sale'. We started by ordering an extra fifty copies of the most popular Sunday papers and these were soon sold out. This led to further supplies from the wholesalers being sent direct to Skegness Station and an even better selling venue being established on the sea front opposite the Clock Tower. The business boomed and on the Bank Holiday Sunday, we sold almost a thousand papers.

I had equipped myself with a money satchel for the Saturday round during which I collected a few tips. These had to be separated from the paper money, which meant that Dad knew exactly the amount. I had a better idea. I bought a second satchel so that I had one slung over each shoulder - that on the left for paper money, that on the right for my tips. During the day both bags became heavier and heavier until I called into a shop and changed coins into notes.

At last it seemed that Dad was successful. He had a profitable business and a potential partner when I left school. Little did I know that tragedy lurked around the corner! It all happened like this. Mum had a telegram from Ipswich announcing Grandad Grayston's serious illness and summoning her to his bedside. She must go at once. But who was to take care of the family in her absence? Dad's brother, Harold, and wife Florrie lived but a few miles away and as Harold was away during the week Florrie came and took over Mum's duties. But it did not end there. Dad and Florrie became involved in an affair.

Harold returned early one Friday evening to find them sharing the same bed. The inevitable row ensued, and voices raised, waking up both Olive and me. Louder and louder they went until all was quiet and the next sound we heard was that of Dad's car starting up and being driven away. Saturday and Sunday were miserable days. Both Mum and Dad were absent and Harold and Florrie were scarcely exchanging words. On Monday Olive and I went to school as usual and after tea went to bed early only to be awakened later with noises downstairs, seemingly moving furniture. In the morning we were forbidden to enter the front room and were told that Dad had died. It was some time later that we learned he had committed suicide in Mexborough, Yorkshire.

The funeral was a few days later but Olive and I were not allowed to attend. All we saw was a horse and cart bearing the coffin as it left the village bound for Willoughby where it was to be buried in un-consecrated ground at the edge of the churchyard. I later learned this was the final disgrace attributed to anyone guilty of the unforgivable sin of taking his/her life. This practise has long since ended.

Chapter Four
The Gables

The Local Authority hearing the news and realising that Olive and I were no longer in a stable home meant that we had to return to the workhouse at Spilsby. Accordingly, a couple of days after the funeral, a car arrived at the garden gate and two uniformed men wearing peak caps emerged. They were thought by us to be police officers but were, in fact, porters come to return us to the workhouse, now re-named "The Gables". We had to say our goodbyes and were broken-hearted believing that we should never see Mum again - a terrifying thought for children of our tender age.

As the porters approached Mum, we were each given a bag containing a few clothes and told to go and get into the car". Tearfully we gave Mum a final kiss and bade her goodbye. Both shattered and terrified as to our future we entered the car and were driven off. Half an hour later we arrived outside a forbidding establishment surrounded by high walls and the only entrance through huge wooden doors. A porter appeared and opened them allowing the car to pass through. We then proceeded to the dark doors at the centre of the main building. The car stopped and we were told to get out.

To my dismay Olive and I were then separated and I was escorted through part of the building, up a flight of stone stairs and into a huge dormitory furnished with thirty to thirty-five beds. Only one was empty which I was told was to be mine. Then I was left alone. Most beds were occupied by old men suffering from either physical or mental conditions. Some were moaning or groaning but appeared to be ignored by the others or the staff. No one seemed to have noticed my arrival and my presence didn't appear to have registered with anyone. Initially I was both frightened and horrified to be placed amongst these sad folk. I had heard of an asylum and, at first, imagined that was where I had landed. But soon I was to learn that this was a workhouse* and was, for a short time, to be my new home. What a depressive life it was during the first few days when I was not allowed to leave the building.

Meals were taken in a vast dining hall and the food served was both poor and tasteless. Between meals I was allowed in a dayroom occupied by six aged ladies, two of whom were confined to wheelchairs. Some of them were disagreeable and made no secret of their objection to our presence. This was the only time that I was able to see Olive but, as there were no children's books or games with which to play, we could only sit and talk, then only very quietly or one of the ladies would bawl out:
"Shut up, we want a bit of peace here."
Fortunately our stay in this awful place was to last only three to four months but they were times forever etched on my memory.

----

*   Workhouse – a public Institution providing accommodation for pauper, unemployable people and homeless children.

Another week elapsed before Olive and I were escorted to Spilsby school and at least one aspect of normal life was resumed. Such was our existence until, three months later when, in the same car and with the same porters our transference to the new Children's Home at Horncastle was effected.

Apparently the Lindsey County Council had changed its policy of dealing with children through the workhouse system. In future all were to be transferred to a new centre to be established at Horncastle; a more suitable environment for children which could be achieved by creating, as far as possible, by placing smaller numbers in individual homes. This resulted in the concept of a "Cottage Home" for twelve and with

The Children and Staff at Horncastle Children's Home 1934
George is shown top right

accommodation for its own foster mother. In pursuit of this policy ten Homes were planned. Sufficient additional land adjacent to the workhouse had been acquired for the ten homes plus house and garage for the Superintendent and several acres for playing fields.

Only four of the new Homes were complete when we arrived and there was no sign of the remaining six. As can be expected the girls were given priority and the four were allocated to them. The boys were less fortunate. They were to live in the old workhouse with very few alterations from the one we had just left. The men's ward was to be known as Home number 5 and the women's ward known as Home number 6.

I was sent to Home 6 to join the thirty or so boys there and later learned that a similar number were in Home 5. Each Home was in the charge of a foster mother with an assistant. Miss Scarborough in the former and 'Ma' James the latter. I quickly observed that the regime was very strict and brutal. 'Ma' James was the most objectionable female one could imagine. Bullying and punishment were her speciality. For the slightest misdemeanour the offender would be attacked with the nearest potential weapon to hand. A hairbrush would be used to thrash the knuckles of the folded hand or, if a coat hanger was handy, it would be used to beat the poor child across the back and shoulders. One ten year old that wet the bed was made to sit in a bath of ice-cold water and there remain for at least half an hour, even longer if 'Ma' James had forgotten that he was there.

The boys ages ranged from four to fourteen, the younger ones placed under the care of the older ones. 'Caring' included washing and dressing and making the bed. All domestic work was done by the boys - beds turned and remade every morning, the dormitory floor polished and all lockers, doors etc. dusted and made ready for inspection at 7.30am. Two boys would be detailed to lay the breakfast table, serve the food and wash up afterwards. Woe betide anyone not on parade at 8.30am for Matron's inspection. This included a visual inspection of hands, face, behind the ears, clothes and shoes. After this it was off to school escorted by at least two members of staff. The main party, of which I was one, was taken to the Cagthorpe School and the remainder carried on to the Church School.

The grey uniform and 'crew cuts' immediately singled us out as 'boys from the Homes' and, initially were not welcomed by the locals. We were made to feel inferior but soon settled in as we proved to be their equals both in the classroom and on the sports field. Names of those near to the top of the class were of our 'kin' whilst others were selected to represent the school in the football and cricket teams. The girls did equally well and found places in the school hockey team.

I enjoyed my time at Cagthorpe and progressed well. The headmaster, Mr G. J. Waymouth was kindly but strict. He suffered badly from the gas warfare of World War 1 and had severe breathing problems but he was never absent. Mr Jenkins was our class teacher and specialised in History and Geography. Mr Hanson taught Maths and Scripture and led football. He also was our referee at all school football matches. Mr Thornton took us for Woodwork and Miss Key for Music. Miss Hemsley taught English. I did well in most subjects though I was weak in Woodwork.

During the last term at school all pupils due to leave were entered in the Jobson Prize Competition. Some years ago the school had benefited from a legacy left by a Mr Jobson which provided £10.00 for the first and second winners. These were for the boys and girls judged to show the most promise of becoming good citizens. The rules of the competition were known only to the teaching staff; 90% of the marks awarded concerned character and behaviour, and 10% were for an essay known as the Jobson Essay.

John Honour

In my year, my pal John Honour won First Prize of three pounds and I came second and won two pounds. When I stepped on to the platform moreover, I was delighted that John had come first and I was more than satisfied to have been a close second. With my two pounds I opened my first Post Office Savings account.

Chapter Five
Starting work

On leaving school in December 1935, unemployment in Horncastle was high with few opportunities for school leavers. Local business folk preferred Horncastle boys rather than those from 'The Homes'. For a few weeks I helped with the stores on the premises which accepted all requirements from the suppliers and rationed them out to the individual homes. All items of food were delivered in bulk: sugar, flour, currants and raisins in sacks and tea in foil-lined tea chests. Each commodity was weighed into 1lb or 2lb bags, placed into stock and recorded in the stock book. Bags were formed using sheets of thick blue paper and folded in such a way as to prevent any leakage. Subsequently, supplies were issued in accordance with orders received from the various Homes. At the end of the year I assisted in checking and recording the remaining stock and calculating its value.

This work, though new to me, I handled well. When checked, my figures were found to be correct and an ambition to become an accountant soon developed. This was never to be realised. Very few vacancies occurred and I did not have a Grammar School education. However as January passed and jobs as errand boys were found for some, others were sent to hotels as kitchen hands, I remained in the stores. At long last I was asked if I was

Roland Achurch

interested in engineering. If so a seven year apprenticeship on low wages would need to be served. Once qualified the reward would be much higher wages than those paid to unskilled workers. I readily accepted the idea and learned that a local firm of agricultural engineers, namely E. Achurch & Sons were offering vacancies for apprentices. I attended an interview with Mr Roland Achurch and was accepted. The starting wage was five shillings a week, with increments of 2s 6d (12.5p) for each of the next four years.The 28th January saw me donned in brown Lybro overalls at the door of the Old Theatre, the tractor workshop. A Mr Frank Carpenter was to be my fitter/tutor. After spending a couple of weeks digging hard baked soil from tractor wheels and washing down vehicles, I began real work.

Frank Carpenter with apprentice George

Chapter Six
Apprenticeship

A great deal of time was involved with repairs to tractors and stationary engines which had either failed to start or had broken down. Tractors always broke down in the field usually whilst ploughing or pulling heavy cultivators, often several fields away from the road, and repairs had to be carried out on the spot.

Our service vehicle was an old Morris Minor van of 1929 vintage with a 'crash' gearbox, cable brakes and six volt lighting. On arrival at the farm, the farmer or the tractor driver would advise us where the broken down vehicle was located and away we would trudge carrying the necessary tools. If in the autumn or winter the weather was wet and cold, and the working conditions far from ideal, whatever the fault it had to be put right. Sometimes it was a gearbox job, probably taking a whole day to dismantle, returning the following day, with the necessary new parts, to reassemble and complete the job.

Many breakdowns were due to engine failure; bearings disintegrated. This involved draining the engine oil, removing the sump and lying on one's back so as to remove the damaged bearing and fit the replacement. Or it might be a starting problem. One had to diagnose first, whether it be a fuel or ignition fault. This done successfully there was still much hard work in store cranking the engine, before it burst into life, there being no electric starters in those days.

Routine services were usually effected with the tractor under cover, often in a cart shed, but if we had the choice, we preferred the relative comfort of the barn. I found the work to be very rewarding even if the farmer seldom said thank you.

In June of my first year the tractor workshop was not very busy and Gordon Spratt, the works foreman, instructed me to join Jack Haresign who used to attend markets and make deliveries. He was to be responsible for setting up the show stand at the Lincolnshire Agricultural Show to be held at Keddington Park near Louth. My first task was to prepare new implements and machines in the yard and help to load them on the lorries. Then there was a nice comfortable twenty mile ride to the Showground before unloading them and positioning them in their appointed place. Finally, all wheels had to be cleaned and metal ones painted with matt black paint. Those with pneumatic tyres, special tyre paint was used and a representative from the appropriate tyre company would colour in the lettering. Great care had to be taken to avoid spillage on the grass which Jack had carefully mown in advance.

This work took the two full days before the show opened. We were lucky with the weather that had been warm and sunny, but during the night it changed; it poured with rain, and for the next two days, it continued. The showground became flooded including the car park. Cars were unable to move or if they did, it wasn't long before they were bogged down. Then George had a bright idea. There was an International

37

T20 crawler tractor on the stand and Jack had taught me how to drive it.

"Jack," I said,

"can I drive the crawler and tow a few cars out of the car park?"

"Alright," he replied,

"but I'll come with you and make sure you're OK."

Proudly I drove to the car park making sure that I had the tow rope with me. First I towed a lorry out and all went well. Then I had a thought. How about making a charge. So I shouted:

"Tanner, for a tow."

Of course everybody responded and for the next two hours I was kept busy. My trouser pockets should have been filled, but farmers being what they were, many escaped without payment. It was still a worthwhile operation.

At the age of fifteen there was an outbreak of diphtheria in several parts of Lincolnshire, Scunthorpe, Market Rasen, Brigg and Horncastle. A mass inoculation campaign was launched but I was one of four local victims who contracted the disease in its most virulent form. We were rushed off to the Isolation Hospital at Scartho, near Grimsby. I well remember the worst symptons being the obstruction of my throat and breathing difficulties. On admission an elderly doctor dressed in 'plus fours' approached my bed wielding a hypodermic syringe in his hands saying to the Sister:

"We'll give him fourteen thousand units to start with and see how he goes."

Sister exposed the left cheek of my posterior and 'bang' I could feel the anti-toxin being pumped in.

At about eleven o'clock the next day the doctor checked the chart at the foot of the bed, examined my throat and said to the Sister:
"He'd better have another fourteen thousand." And repeated the previous day's performance. I was still very ill and there appeared to be little improvement. On the third day he returned and reduced the dosage to seven thousand units and my temperature began to fall. That seemed to be the end of the treatment and progress, though slow, continued. That was until the tenth day when I came out in an awful rash, covering my whole body. All patients suffered the same rash that itched like hell. Sister seemed unconcerned, even made a joke of our discomfort and promptly issued each one of us with a large jar of green jelly.

"Just apply this wherever it may start itching," she instructed. We followed her orders and liberally applied the aforementioned jelly continuously until it penetrated pyjamas, sheets, and the lot. It took about ten days for the itching to subside and I was beginning to feel very much better. The next procedure was to have throat swabs taken daily as three consecutive negatives were necessary before one could be declared free of the disease and ready for discharge. The problem was that the pattern of swabs results in my case went one negative and one positive or two negatives and one positive and it wasn't until I had spent thirteen weeks at Scartho that I had my three negatives and was allowed to be discharged.

All children had to leave The Homes by the time they were sixteen years of age. Many were placed in situations which included accommodation but for those employed locally alternative arrangements had to be made. Bearing in mind that I was among the first batch to be dealt with, it would appear that whoever was faced with this problem was doing it for the first time. The selection of 'digs' left much to be desired. Perhaps it was done by contract. I know not. I can only relate the events.

Four of us were sent to a Mrs Brown at 63a West Street, Horncastle and shared a small bedroom. The food was awful. The Brown family had their evening meal at 5pm. Our meal was at 6pm. All signs of the previous meal was cleared away and, if we were lucky, we were served with a watery soup followed by bread and cheese. Occasionally we had a poached egg on toast. It was obvious that this was to be a low cost tariff. Had the authorities taken the lowest priced contract? Furthermore, it was made perfectly clear to us that our presence was not welcome in the evenings, and at weekends we were expected to disappear.

I was not the only one unhappy with the arrangement but complaints to the Superintendent were fruitless. I talked to my work mate, Frank, who was in digs only fifty yards from the workshop, and spoke very highly of Mrs Green, his landlady. She charged £1. 2 shillings a week but I was earning only twelve to fifteen shillings including overtime. No doubt our current 'digs' were subsidised. We handed our wages in to The Homes each week and were given 1 shilling back as pocket money.

Frank approached Mrs Green on this matter and she eventually agreed to take me when a vacancy occurred, at the rate of £1.00 per week until such time as my earnings would allow me to pay the normal amount. Several months later I made the move. I shared a bedroom with one other fellow and the food was excellent and in sufficient quantity as to permit a second helping if required. It was obvious to me now that Mrs Brown had been in it only for the money.

There still remained the problem of the evenings. Landladies prefer to have the privacy of their homes during the evenings and weekends. For the first year with Mrs Green, I stayed in and pegged a hearthrug for her in appreciation of my not paying the original charge, and this was appreciated. But spare time was still a problem. It was at this time I met Ted Byron, the Secretary of The Horncastle War Memorial Hospital. He was also a newspaper reporter for the Lincolnshire Echo. Being the secretary of the local branch of Toc H he invited me to their meetings.

For many years Ted became a valued friend and extended much kindness to me, particularly when I was a teenager and my needs were greatest. He seemed always to be involved in organising functions in the town of one sort or another - whist drives for St Dustan's Blind Society, the annual Fete in aid of the War Memorial Hospital etc. and this gave me the opportunity to lend a helping hand.

Ted was in digs with the Crowder's, farmers at Thimbleby, just two miles from Horncastle. His home, to which he returned, at the weekends was in Lincoln and many a time he invited me to accompany him. His ageing parents, referred to as "Pop" and "Ma" always welcomed me and I soon learned they loved a game of cribbage in the evening. This almost became a ritual and many hours, often long after midnight, were spent in battles as to which pair were the winners.

Chapter Seven
Toc H and the Town Band

In December 1938 a group of members of the Horncastle Branch of Toc H travelled to London to attend the 21st Birthday Festival. Padre H. P. Laurance drove us there and the party included J. Hall, the chairman, Ted Byron, the secretary, John Honour and me. On reaching Yaxley, near Peterborough, one of the rear wheels of our car picked up a puncture but fortunately we were near a garage with a fish and chip shop next door. A mechanic mended the puncture whilst we indulged in lunch after which we continued our journey.

On arrival we checked in at our hotel, had high tea and then attended a Thanksgiving Service before proceeding to the Albert Hall. This was my first visit to London and I was thrilled to bits particularly as I was to take part in the festivities. During the programme, the Albert Hall was plunged into complete darkness with the exception of the single flame of an oil lamp bearing the double cross of Lorraine and designated "The Prince of Wales Lamp."
Then two spotlights focussed on the appropriate stairways from the gallery to illuminate the Lamp and Banner of each Branch. These proceeded down the stairs to be re-kindled from the Prince of Wales Lamp followed by others from Branches across the world until all Lamps had been re-kindled.

I was very proud to be the Banner Bearer for the Horncastle Branch with Ted Byron carrying the Lamp.

The whole ceremony was in complete silence and with the only illumination from the flickering Lamps a most impressive one. After the Festival we returned to the hotel and being exhausted, retired for the night.

On Sunday morning we attended Holy Communion at All Hallows Berking by the Tower officiated by "Tubby" Clayton, the Founder of Toc H. Then followed a conducted tour on an open charabanc and lunch. In the afternoon there was a gathering of all the members in the Seymour Hall for a friendly chat from "Tubby" sitting on the floor with his black spaniel dog at his side. Thus ended an event which ever lives in my memory.

At the introduction of food rationing in 1939 Ted was appointed Food Officer for the Horncastle area, his first task being to set up the necessary administration and issue Ration Books to every person resident. This was an enormous job as each book had to bear the name, address and registration number of the recipient. There were no computers in those days. All had to be hand-written and many volunteers were recruited of which I was one. So it was that my friendship with Ted, which was to endure over forty years, developed in the early years of World War II.

It was at the Toc H meetings I met for the first time local businessmen and members of the teaching profession who, to my surprise, were friendly towards me and seemed to treat me as an equal. I subsequently became a member after serving twelve months as a probationer. I was impressed with the wording of the

Toc H Compass which was - To think fairly, To witness humbly, To act wisely and To build bravely. A number of members were deputed to give special attention to new members, particularly to young lads who included those from The Homes.

Several of my contemporaries joined Toc H but at about the same time there was a recruiting drive by the Territorial Army to swell their ranks. The clouds of war were threatening over Europe once again. Our former Prime Minister, Neville Chamberlain, may have appeased Hitler over Czechoslovakia but the necessary preparations for war must be made. A military brass band marched down the Horncastle streets on a certain Thursday evening which corresponded with the monthly meeting of Toc H. The invited speaker was none other than Captain (now Major) Allbones, previously our choirmaster at St. Mary's Church. He made a strong appeal to all those of military age eighteen to fifty, and urged them to enlist with the Territorial Army. They were to be known as the Militia. Many responded to his call including several old school pals of mine. When war broke out in September 1939, they were the first to be called up forming the core of the 4th Battalion, Lincolnshire Regiment. There were only few of my age left in the town and they were either in reserved occupations or awaiting their call-up papers.

For me these were both lonely and unhappy days. Apart from overtime at work I had nothing to do. I did a bit of voluntary work and was then asked if I might be interested in joining the ARP services. That's how I became first a messenger and then, after a special course, a fully qualified member of the decontamination squad.

At about the same time it was made known that the Town Band was decimated by virtue of the fact there were so many players now in the armed forces and the Band was seeking new members. They needed a bass drum player and this appealed to me so I offered my services and was accepted. Band practise was on Sunday afternoon and there was either a concert or parade to attend most weekends. Thus I once again became fully occupied. But the war was not going well. Hitler's armies swept across Europe and concluded with the Fall of France and the Dunkirk Evacuation. Then came the Battle of Britain, news filtering through of some of my ex-school pals being killed in action.

Chapter Eight
Enlisting & Overseas Posting

The time had come for me to reconsider my situation. Although I was in a reserved occupation I could volunteer. I had no relatives or dependants and no one

would miss me or be hurt if I failed to return. And so it was that one Saturday morning I made my way to Newport Barracks in Lincoln and enlisted in the RAF Volunteer Reserve. I must confess, however, that all my intentions were not altruistic. I had other objectives in mind. If I could join the engineering section of the RAF and survived the war, the special training would serve me well in the future pursuit of my career in

George enlisted 31-1-1941   the agricultural trade.

I was accepted, signed on the proverbial line and was issued with a railway warrant as I was destined for a huge recruitment camp at Padgate. Four days there and after medical and trade tests I was issued with my RAF uniform and posted off to Morecambe for initial training. This included 'square bashing' on the promenade, arms drill and an abundance of 'jabs' (inoculations). Six weeks later a passing out parade took place and I was posted to Kirkham, between Preston and Blackpool for the necessary technical training.

This establishment was known as the 10 S of TT which, given its full title read 10th School of Technical

Initial training at Morecambe

Training and served the purpose of training both aircraft engine and airframe mechanics. In addition trainees were used for guard duty, fire picket and hard to believe, parachute patrol. In order to be effective at this duty the 'powers that be' armed us with a broom handle and instructed us to keep our eyes sky-wards and raise an alarm - by shouting "Invasion" if we saw any signs of a parachute.

After the Flight Mechanics course of thirty-eight weeks, a 14 day leave pass was awarded and, as I had exceeded 70% marks in oral, written and practical examinations, I was instructed to return and take a fitters course of ten weeks duration, before being posted to an operational unit.

I started as an AC2 Under Training (UT) until such time as I had completed the Flight Mechanics Course.

I then became a fully fledged AC2 and the daily pay rate was three shillings increased by three pence. On completion of the ten week Fitters Course I was reclassified as an AC1 with a further three pence a day increase. Eighteen months later I was promoted to Leading Air Craftsman with an additional three pence and as I was overseas, it entitled me to a further sixpence overseas pay, giving a total of 4s 3d per day.

On completion of the course I was posted to 106 Squadron, stationed at Coningsby, only eight miles from Horncastle. This Squadron was flying Handley Page, Hampden bombers fitted with Bristol Pegasus radial engines. It mattered not that I had been trained on Rolls Merlin engines. Such was the wisdom of those responsible for postings.

I was attached to a maintenance gang of fitters with whom I got on well. Most of our time was spent carrying out routine fifty and one hundred hours inspections and correcting any malfunctions that occurred. However, one morning in November 1941, the tannoy address blared out:
"Attention all personnel. In one weeks time this station is to be visited by a very important person and I shall expect everywhere to appear in perfect condition. Your Squadron and Section Commanders will be issuing the necessary instructions."

All flying operations were cancelled. Hangar floors were cleaned using one hundred octane fuel: paint and whitewash was applied left, right and centre and the station was made fit for even a king to see. Little did we know, until the appointed day, that it was to be none other than King George VI. On the very day he should have arrived at 11am fog delayed him and it was not until 4pm that his car arrived. The King entered No. 1 hangar where the aircrew of 97 Squadron were assembled and he presented them with various medals, then did the same to 106 Squadron aircrew in No. 2 hanger before leaving. The ground crews and other staff only had a distant view of the royal party before the visitors were whisked away into the fog. It mattered not that we had been on parade since 7.30 am without even a meal or comfort break.

My sojourn at Coningsby was to be short lived and in April 1942 I was posted overseas. First I went to West Kirby, near Liverpool, to be fitted out with tropical kit and then to board a troopship, The *Highland Brigade*. She was a refrigerated ship and in peacetime regularly conveyed meat from Argentina. We sailed up to Gairloch and joined a convoy. Sailing due west we hoped our destination, still secret, would be Canada but after three days we steered south. An Atlantic gale blew up and the ship pitched and tossed with devastating effect on most personnel aboard. I was particularly fortunate for although I had never been at sea hitherto, I was unaffected. The convoy speed was said to be four knots and as we were following a zigzag course our progress was slow. Eventually we turned east and in due time arrived at Freetown. Then, for some unknown reason we were transferred to the sister ship named the *Highland Monarch* and proceeded to Cape Town.

There we disembarked and boarded the *Niew Amsterdam*, a Dutch liner being temporarily used as a transit camp. This was due to the fact that the Imperial Services Camp at Polesmoor was flooded following heavy rain. After seven days aboard and being allowed shore leave each afternoon, we transferred to Polesmoor for a further fourteen days. The rumour was that enemy submarines were attacking convoys from the Mombasique Straits.

The next part of our voyage was diabolical. We boarded *RMS Scythia*, allegedly rescued from the scrapyard. Capable of carrying a mixed cargo and five hundred passengers, four thousand troops were taken aboard and she then sailed to Durban. There a further number of troops plus three hundred nurses and officers boarded her and then sailed east. I was on "H" deck and meals were served on "B" deck individually. This meant negotiating four or five flights of stairs on the return to "H" deck. Just imagine carrying two plates of food, one of which was of a watery stew, down these stairs with well worn metal edges. Inevitably spillage occurred and first one and then another slipped. They went flying and in the process, lost their entire meal, forming a treacherous surface for others to follow. It wasn't long before all five flights were similarly affected. No wonder the troops reacted and a near riot situation developed.

The second day out the Officer Commanding Troops announced that we had fourteen days food and water aboard but that we should not see port for eighteen days. This cheered no one particularly as rumour had it that we were destined for Singapore. This proved wrong as our course changed to north and eventually, we arrived at Aden, but only to take on water.

Our voyage in a northerly direction, took us through the Red Sea where temperatures were up to 100 degrees Fahrenheit and even higher below deck. Drinking water throughout the voyage was turned on twice a day and then only for a period of one hour. Conditions almost reached riot proportions and the troops were only calmed down when a naval chaplain appealed to the O.C. Troops to divert food destined for the officer's mess to the troop's galley below. A good meal was then served to all troops and the situation was saved for we had already been warned that every one would be batonned down below and an armed party of Royal Marines would come aboard and fire, if necessary, on any individual or group seen to be leaders of any disorder.

Certainly, this calmed down the atmosphere and for the last few days of the voyage, there was no further trouble. Every one aboard must have been relieved when *RMS Scythia* moored in Port Tewfig (now spelt Taufig) harbour. All personnel disembarked, the RAF troops being directed to a waiting train in a nearby siding. For the next couple of hours we travelled westwards, passing several military establishments en route until we stopped near a very large camp, known as RAF Kasfereet. With a capacity of ten thousand comprising both ground and air crew, it was the major assembly location for the whole of Middle East Command.

Tents were arranged in rows which seemed to stretch for miles with very few unoccupied. But a few of our draft found empty ones. The remainder slept, that first night, under the stars. And here we learned our first lesson. The ground sheet we carried went on top of our blankets otherwise, with heavy overnight dews, they would be soaked long before dawn. The following day loads of tents arrived which were quickly assembled in equally long lines matching the pattern of the others.

There were, supposedly, parades every morning at 9am when the postings were announced but, since there were no loudspeakers within hearing distance, only a few hundred bothered to amble out of the tent. It was ten days before an announcement asking for engine fitters to volunteer for a temporary attachment to a local civilian airport was made. I was fed up hanging around with nothing to do, so I volunteered. A group of us reported to the Guardroom where transport awaited us and we were whisked off to the British Airways Repair Unit at Heliopolis.

Whilst on detachment there I applied for ten days leave in order to take advantage of a travel offer being made by the YMCA. This was a trip up the River Nile as far as Luxor, Thebes, the Valley of the Kings and the tomb of Tutankhamun at the bargain tariff of £11. Together with five others the leave was granted and off we went. The outward journey up the Nile was by rail, albeit in carriages with wooden seats. The first visit was to the Temples of Luxor and Karnak. I was absolutely astonished at their immensity, the height of the columns and their detailed carvings and decoration.

When we reached the Valley of the Kings and entered the Tomb of Tutankhamun I was amazed, on arrival in the burial chamber, at the contents which had been stored in such a confined space. These are now in the Museum in Cairo.

This was my first sight of hieroglyphics, the pictorial language used by the ancient Egyptians, to record the life history of the king who died at the age of eighteen. We returned down the Nile on a 'Felucca', a particular type of boat, sail powered and used mainly for transporting cargo. This was a most educational and enjoyable trip which was etched for ever in my memory. Furthermore, my appetite for further travel in future years was now seriously stimulated. I was then reminded that Sagittarians are supposed to be dedicated travellers and I was 'hooked'.

On my return to the British Airways Repair Unit at Heliopolis, for the next sixteen days, we carried out routine services on aircraft in transit. I, together with three others were then recalled to Kasfereet and posted to an unknown destination.

Chapter Nine
Malta and the Siege

That night we boarded a Dakota C47 and flew to Malta. Our landing was delayed as a bombing attack was in progress but half an hour later we landed on a runway which had been cratered during the raid and only temporarily repaired. As we left the aircraft the sirens sounded again which necessitated our being hurriedly directed to the nearest air raid shelter. After this raid we were conducted to Luqa Poorhouse where we slept the remainder of the night.

The next day I was posted to 1435 Squadron Maintenance Flight who were responsible for the fifty and one hundred hour inspections on the Squadron's Spitfires and changing the engines when required. I joined a gang of two headed by a Corporal Harry Marsden who, before being called up, was Manager of one of Woolworth's Stores. With but a few hand tools and one 'Rolls Kit', to be shared between "A", "B" and "C" Flights and Maintenance Flight, one could not imagine us to be over equipped. Added to our difficulties was the lack of spare parts. Our only source was to cannibalise crashed aircraft of what serviceable parts remained. So often you hear 'necessity was the mother of invention' and it was certainly so during the siege.

The island of Malta was in a state of siege from May 1940 to September 1943. Much has already been written in the plethora of books which have appeared in the post-war years. I shall content myself by mentioning the role played by the ground crews whose duty it was to keep the aircraft flying. Inevitably, there was a great deal of improvisation. In Maintenance Flight we had no such thing as a 'plug cleaner' or a 'torque spanner', both of which were necessary when spark plugs were cleaned and tested according to regulations. So, after studying the text book known as the AB 1464, and scrounging scrap materials from one place and another, both items were produced and put into service. Since there were up to eighteen aircraft on the Squadron, each Merlin engine having twenty-four plugs to be changed every twenty-five flying hours, it will be appreciated that improvisation more than paid off. Ground staff worked every daylight hour irrespective of continued bombardment. The only time we sought shelter was after a red flag had been raised indicating that a local attack was imminent. Our service area was on Safi Strip on the edge of Luqa airfield where both fighter and bomber aircraft were parked – one of the most blitzed areas on the Island.

In World War 11 Malta was the lynchpin of the war in Europe. Situated almost a thousand miles from both Alexandria and Gibraltar and the enemy air forces less than sixty miles away it was inevitable that Malta would be heavily attacked. General Rommel with his German Afrika Korps occupied North Africa and relied on supplies from Europe. The Allied Armies depended on maintaining Malta so as to keep open our routes to the Middle and Far East.

In May 1940 the Italian Dictator Benito Mussolini believing that the German armies under Adolf Hitler had all but won the war in Europe, allied the Italian nation to the German cause thus forming the "Axis Forces".

On the same day the Italian air force, the Regia Aeronautica, made its first bombing attack on Malta for if the island was to be occupied there would be no threat to the supplies to Rommel's armies in Africa. On the other hand whilst the British could hold out, the route to the east would be maintained and the enemy could be attacked from both sea and air. Thus began the most critical siege of the island in history.

Sadly the RAF had no modern fighter or bomber aircraft apart from a few Gloster Gladiator biplanes still in their packing cases. Four of these were hastily assembled and flew in combat destroying a number of the Italian heavy bombers. Some weeks later the first of the Hurricanes arrived but were heavily outnumbered by the Italian planes based in Sicily only fifty four miles away. Air raids continued night and day in spite of the heavy and light anti-aircraft defences.

Spitfire from 1435 Squadron

The first small consignment of Mark V Spitfires arrived and took up combat with the enemy. Later the Aircraft Carrier *"Eagle"* delivered Mark IX Spitfires to within flying distance off Malta. On arrival in Luqa and having to be serviced, fuelled and armed, the delay left them vulnerable to the German Air force who promptly bombed and destroyed most of them. The next consignment of Mark IX's were previously serviced and armed whilst in transit on the aircraft carrier. All that was necessary was refuelling at Luqa on arrival. Soon after landing and being refuelled the enemy bombers were shocked to meet them in combat so quickly and suffered heavy losses.

Enormous areas around Valletta, the docks and airfields had been blitzed and demolished. During that time Malta was bombed night and day. The towns of Floriana and the Three Cities - Senglea, Conspicua and Victoriosa were completely destroyed. The Docks and airfields were continuously bombed. Runways at the airfields were frequently bombed and as soon as the craters occurred, a gang of soldiers with a lorry load of stone would immediately fill them in and a "Valentine" tank of WW1 vintage would trundle up and down in an effort to level the site. All this transpired while the air-raid continued.

Supplies of food, fuel and ammunition had to be shipped from the UK., by sea. The convoys were subject to unmerciful attacks once they had passed through the Straits of Gibraltar and entered the Mediterranean Sea. Losses of our shipping were frightening as were both Royal Navy and Merchant Navy personnel.

Supplies in Malta were so low, from 1941 onwards week by week the "Surrender Date" had to be calculated. Food was severely rationed to the troops and civilians alike. All public transport was suspended. No petrol was allowed for private cars. Even Lord Gort, Governor of Malta and Air Marshal Keith Park, Air Officer Commanding, were seen to be cycling around the island.

Our feeding arrangements were as follows, breakfast was served at 6 am., for over two-thousand service-men. We were often made to disperse due to an air raid when a couple of ME 109's appeared dropping their load of bombs at will. The queue had to reform which meant losing your original place. Breakfast was then served and we proceeded to the service area on Safi Strip. Lunch was served (such as it was) from a bashed and battered Bedford Bus with part of one side removed and modified to be utilised as a servery. This was affectionately known as the "Mungy Wagon" which circumnavigated the airfield serving the small groups of ground crews on its way. Evening meals were served in the same manner.

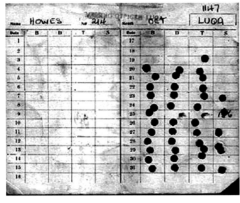

Ration Card issued to 214 Hows

For feeding purposes each person was issued with a ration card which was punched by a tyrannical corporal Jones, who was suspicious of anyone who was trying to beat the system. Some "smart Alec" would apply oil to the card by rubbing it firmly in with their thumbs, causing the hole punch system to be somewhat ineffective and enabling it to be used for a second helping. Such were the desperate methods employed at the time of such meagre rations.

A staggering fact known to few is that more bombs fell on Malta in April 1942 than upon London during the worst year of the Blitz. But this was not the only problem. Food, fuel and ammunition were in desperately short supply and had to be shipped from the UK. Convoys were heavily attacked from the air and sea and a further hazard was the German and Italian submarines. In fact, without the convoys Malta could not have survived. In August 1943 the Island was almost starved out. Fuel was down to the last dregs and the 'Heavy Ack-Ack' crews, many manned by the Maltese gunners, were limited to firing only a few rounds each day. Later we learned that Malta was within ten days of surrendering.

In April 1942 the Island of Malta was awarded the George Cross by His Majesty King George VI in recognition of the gallantry and sacrifice of the Maltese subjects and members of

The casting of the commemorative Bell to mark the Siege of Malta

60

H M Forces involved in the defence and supply during the Siege.

After the war the George Cross Island Association (GCIA) was formed and the Siege Bell* Memorial was erected in the lower Baracca Gardens, facing the Grand Harbour overlooking Valetta. In April 1992, Her Majesty Queen Elizabeth II attended the inauguration of this memorial. This was timed to commemorate the fiftieth anniversary of the award of the George Cross.

GCIA Veterans continue to attend annually a memorial Service held on the site.

The Midlands Branch of the GCIA of which I am a member meet regularly at the Senior Services Club in Leicester, Graffham Water and High Wycombe.

----
*A reproduction of a five hundred year old twelve ton bell was selected to sound daily across the Grand Harbour in Valletta. This was cast by the Bell Foundry Taylor's of Loughborough. A replica of the 'Glorioso' bell in Erfurt Cathedral northern Germany and together with a reclining bronze figure representing the people of Malta who paid the supreme sacrifice in World War 11, completes The Siege Bell Memorial.

Chapter Ten
The Pedestal Convoy

In July 1942 Winston Churchill is recorded as having ordered a heavily protected convoy to be assembled immediately to relieve the Siege. Fourteen merchant ships accompanied by no less than forty escorts including four aircraft carriers, two battleships, seven cruisers, twenty four destroyers, two tankers, four corvettes and eight submarines left these shores and reached Gibraltar without incident. The name of the convoy was "Pedestal". The enemy "Axis Forces" met the convoy with a formidable force comprising: sixteen Italian and five German submarines which attacked the convoy between the Straits of Algiers and the Sicilian Channel supported by seven hundred aircraft; eighteen Motor Torpedo boats lurked between Cape Bon and Pantelleria.

This ship was torpedoed, bombed and rammed by a plane. She made it, and so did Malta

Nine merchantmen, one aircraft carrier, two cruisers and one destroyer were sunk in the next five days onslaught. Four more merchantmen were badly damaged and had to leave the convoy. The ordeal of the tanker "Ohio" became a legend. An exploding Stuka crashed on her

The relief of Malta by the Pedestal Convoy
August 1943

decks; her boilers blew up and the engines failed; she was twice abandoned and twice re-boarded. With decks awash, she was lashed between two destroyers and led into Grand Harbour, Valletta. The previous day four merchantmen arrived and the Siege ended.

Subsequently Rommel and his Afrika Korps were being driven out of Africa and Malta was to become the base for the invasion of Sicily.

Food supplies were desperately short and rations were reduced to one third of normal and, as happens in such dire situations, a 'black market' thrived. For example, at Quormi, a village adjoining Safi Strip, an old lady sold pancakes made from black market flour and water mixed in a disused 7lb jam tin and cooked over a smoky fire in a frying-pan. As no form of cooking fat was to hand, anti-freeze oil from the camp became the substitute. Though tainted with the taste of paraffin at least Ma's pancakes partly filled the empty stomachs.

Her business thrived and at 3s 6d each, equal to a day's pay, she must have made a huge profit, even though she claimed to have paid £1.00 for every 1lb. of flour. Service personnel had provided anti-freeze oil which was used for cooking.

Not all our aircraft were lost in enemy action. Sadly some perished as a result of accidents. Here is an example. All aircraft at Luqa were parked in blast proof pens constructed of empty four gallon cans filled with earth and rocks and stacked so as to form walls eight cans wide at the base and three wide at the top. The overall height was ten to fourteen feet according to the aircraft that was to be parked. Our Service Pen was numbered 125 and was on Safi Strip. Beyond were Pens for 'kites' in transit to the Middle East that had called in for refuelling, and others that were operating from Malta.

Christmas Card locally produced during the seige of Malta G.C. 1940-43

One afternoon whilst servicing a Spitfire, a voice was heard shouting:
"For Christ's sake evacuate." And seeing a column of smoke emerging from a pen reserved for a Wellington aircraft, we ran like hell in the opposite direction. This led us back to Luqa School where we were billeted. Word soon got round that an armourer had tripped while clambering over the main spar and had dropped the flares he was carrying. These were normally fused after they had been placed in the flare tubes on the

aircraft. On this occasion, and against strict rules laid down, the flares were fused in the Armoury workshop. Thus as soon as they were dropped they lit and set the fire going.

The situation was very much worsened as an aerial torpedo was on board and it was only a matter of time before a massive explosion would occur. Some thirty minutes later the explosion took place, and seconds on, fragments of an engine showered down in the playground. Fortunately there were no casualties but the offending armourer was court-martialled and awarded eighteen months field punishment for the offence.

Needless to say no flares were fused thereafter until safely installed in the flare tubes. Another result of the siege was the lack of communications with our families at home. For a period of over twelve months no surface

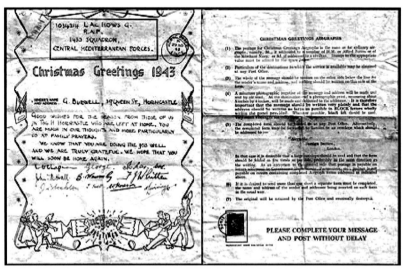

Airgraph from Toc H Branch Horncastle

mail or parcels were despatched from the UK. Only the occasional air mail letter got through.

A system of Airgraphs was introduced which involved writing a letter on a special form which was then photographed, reduced on micro film, despatched by air and reproduced on arrival before being sent on to the intended recipient. But in October 1942, a load of surface mail, which had been lying in Liverpool docks over twelve months, did arrive. A notice was issued and all personnel were invited to search for their own mail in a room at Luqa HQ. This room was about twelve feet by eight and mail was stacked to ceiling height. Searching for ones own mail was worse than looking for a needle in a haystack. Mail included, was for those posted home or who were killed in action. After three weeks it was realised that much of the mail would never be claimed, and finally it was removed and destroyed.

The first photograph of Joyce received in Malta

However, after several searches I was fortunate to find a few letters and two newspapers, some of which had been addressed to an APO (Army Post Office) in Egypt and forwarded to Malta which, by then used a different APO, now designated CMF. (Central Mediterranean Forces). Amongst the mail I retrieved was a letter that was to change my life for the next fifty years. So here is another story. One letter was in a familiar handwriting. It was mother's, who now took the name of Aunty Ethel, and enclosed with the letter was a photograph of her niece, the daughter of her sister Nellie. It was explained that the picture was of Joyce, now training to be a nurse, who was anxious to have a pen friend serving in the Forces abroad. Could I be interested? Of course, I was. At the first opportunity I wrote to Joyce and to my surprise a letter from her arrived three weeks later.

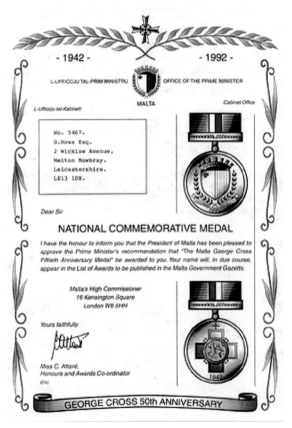

The George Cross
commemorative medal awarded 1992

Chapter Eleven
The Italian Campaign

By now the German Armies had been driven out of Africa and on July 10th 1943 the Allies attacked the enemy in Sicily. Within a month the island had been recaptured and, in September, the invasion of Southern Italy commenced with landings at Salerno. Later 1435 Squadron moved from Malta and aboard Landing Craft Transport, (known as LCT's) crossed the Mediterranean to Taranto, on the mainland.

On arrival we were transported to Grotagglia some ten miles away, where a fair weather landing strip had been prepared. These strips were constructed on levelled land where the soil had been compacted, so could be quickly prepared and made suitable for fighter aircraft to operate. The role of 1435 Squadron was to intercept enemy aircraft which might attack convoys from Taranto.

A tented camp was established in an olive grove, but the fact that it was low lying land at the end of the runway had been overlooked. On the third night, when we all settled in bed, a violent thunderstorm broke out and heavy rain flooded the camp. Fortunately for me, I had raised my bed eighteen inches above ground level by acquiring four stakes and a wooden door from a derelict farm nearby. Whilst others made an emergency evacuation I stayed on observing a river of storm water passing beneath me.

Next day the camp was moved to higher ground but as there was no enemy action during the next ten days, we moved on to Brindisi on the Adriatic coast. Here was a harbour, a permanent air and seaplane base with all the necessary associated buildings including hangars,

Blitzed Hangers at Brindisi

permanent and temporary buildings with mains electricity and other services. Ground crews moved into wooden huts fully equipped and aircrews into fully furnished brick built barrack blocks.

The cookhouse arrangements were based on the 'field kitchen' that involved cooking outdoors. Food was placed in metal containers and heated over a fire, burning whatever fuel was available. When supplies of solid fuel, either coal, coke or wood were exhausted, an alternative had to be found. In our case an ingenious device made use of waste engine oil and consisted of two five gallon drums, one filled with oil and the other with water. Each was fitted with a tap and a length of copper pipe leading to the site of the fire - a metal plate which was pre-heated by placing a petrol soaked rag and lit. When hot a drop of oil was allowed to fall on to the plate and began to burn. It was then aided by a drop of water falling into the centre of the preceding drop of oil, dispersing it and aiding combustion. This was known as the 'spit and fart' system, was messy and produced clouds of black smoke.

I had a better idea. Why not use petrol as the main fuel? There was 'oodles' of aviation petrol about and I had studied the principles of a blowlamp as described in the AB1464 (RAF Engineers Manual). By making a downward coil at the end of the pipe leading from the fuel container and bending the end upwards in the centre of the coil the initial flame heated the coil and resulted in a pressurised flame, as in a blowlamp. More heat was generated and there was no black smoke or fumes. The news soon reached the ears of the Engineering Officer and I was severely reprimanded. Misuse of RAF property, i.e. the petrol, was a courtmartial offence. Hows was in trouble and the cookhouse had to return to "spit and fart".

But I was not to be thwarted and wandering round a nearby farm, I located a six hundred gallon tank almost full of diesel. Experimenting with various shapes of burners made out of 1 inch bore pipe and drilling small holes at intervals to form burner jets I succeeded in making an efficient burner. My first attempt was to dig a trench long enough to accommodate up to six food containers and cover it with a length of Somerfield Track (pressed steel plate used for temporary roads). A chimney constructed of several five gallon drums welded together, ends having been removed, and supported by lengths of telephone wire, completed the 'Hows Patent'.

As the food in each container was cooked it was replaced with one filled with water. Thus 1435 Squadron Airman's Mess had the most efficient cooking in Brindisi.

I returned to the Airman's Mess and installed my new contraption to the delight of the Sergeant Cook, a Cockney. Again the Engineering Officer appeared on the scene, congratulated me on my ingenuity and ordered me to make one for the Officers Mess. This was a lovely mansion on the outskirts of Brindisi and the usual 'field kitchen' was in use. I completed the installation in a couple of days and the Sergeant Cook at the Officers Mess was instructed as to its operation. All went well for a week until one evening the Engineering Officer tore into the billet in a panic.

"Hows," he yelled, "there's been an explosion with your b....y contraption. Get down there at once. We don't look like getting any dinner tonight."

I rode in the Jeep with him to the scene and could scarcely believe my eyes. Indeed there had been an explosion. My well constructed chimney stood awry and bent into a 45 degree angle, the fire was out and cooking containers were scattered around.
"Where's the cook," I enquired.
He was summoned and I asked him,

"What on earth happened?".

"Well" he said, "I took the first container off when the food was cooked and as there wasn't a spare one about, I blanked the space off with a jerrycan. I saw that it seemed to begin bulging and then there was a hell of a bang. I ran away as fast as I could and when I returned I found that half of it had gone through a bedroom window and the other half was in that apple tree."

On inspection I discovered that the filler cap was still closed and deduced that a small amount of petrol must have remained in the can when it was placed in position. Overhearing the evidence I was exonerated from blame. The Engineering Officer rounded on the poor old cook and bellowed:

"You b....y fool! Don't ever put a jerrycan on there again." Then turning to me he enquired:

"How soon to get it going again, Hows?"

"Within the hour," I replied, and this was accordingly done.

At Brindisi washing facilities were somewhat basic with no provision for bath or shower so I felt it was time for action. Exploring a derelict farm next to the camp, I walked into the cowshed and spotted a vacuum pipeline that had been used for machine -milking. This gave me the idea of converting it into a shower facility. Using a discarded forty gallon barrel filled with water, coupling one end of the pipe to the barrel and the other to the vacuum pipeline and employing a semi-rotary pump (scrounged from the Americans) we had the basis of a unit for six men to shower simultaneously. Sprinklers were made from porridge cans (American) pierced with an ice-pick and fitted to each pipeline outlet.

My invention worked well and was applauded by all during the hot weather, but further modifications were necessary for winter use. A second barrel was employed, an oil burner placed beneath and, using a two-way valve robbed from a wrecked aircraft a hot shower resulted (yet another Hows invention). This was the best accommodation we experienced since leaving the UK.

Cyril Edgeley
Technical Assistant

This was the time when the shortage of manpower in the Services was such as to extend conscription to the 'over fifty' age group. Men were enlisted to serve in the unskilled duties. After the initial six weeks of 'square bashing', they were posted direct to operational units either in the UK or overseas. One of these recruits was sent to our Squadron whilst stationed at Brindisi. He was Cyril Edgeley, a wholesale grocery salesman in 'civvy street'. The Engineering Officer suggested that I might be able to utilise him as a fitter's assistant. At the time I was a self appointed 'spare parts and plug king' He was just what I needed. I could train him to clean and service spark plugs which would become a full time job and give more time for me to chase around cannibalising bits and pieces off wrecked aircraft. This job suited him and within a week or so he became proficient and worked well. As time went by we became great friends and he stayed with the squadron until we eventually disbanded on VE Day.

A huge American ship anchored in Brindisi harbour. Fighter aircraft with their wings folded upwards were the deck cargo as well as a fleet of Jeeps with trailers and the accompanying personnel appeared as or when they had disembarked. We were astonished as we observed the passing convoy. Each Jeep had four passengers, and the trailers contained a square tent complete with wooden floor, fireplace, fuel and personal kit.

About fifty went by followed by a series of air-conditioned self-propelled mobile workshops, over fifty feet long. All the crews from the vehicles set up camp on the far side of the airfield and later that day, the aircraft were unloaded on the quay side next to the hangars. They were huge machines the largest fighter built in the United States called the Thunderbolt. The American crews rapidly assembled and serviced them before appearing on the runway when, with a mighty roar, the planes took off and flew down to Lecce, some sixty miles south of Brindisi and the Headquarters of the 15th Bomber Group, United States Army Air Force (USAAF). They were to be the escorts for the Flying Fortresses and Liberators inflicting daylight raids on Central Europe, their main targets being the Ploesti oil fields in Rumania.
Our Spitfires were to intercept any enemy aircraft

threatening the airspace within a fifty mile radius of Brindisi. In the early days after our arrival several 'nuisance raids' were made by Messerschmitt 109's and Foch Wolfe 190's fighter bombers, but they were engaged by our Spitfires and shot down or turned back and jettisoned their bombs at sea. One day our air space was 'jumped' by four Me 109's flying at 15,000 feet. As soon as they were detected four of our Spitfires took off, climbed above them, shot down all of them and landed within twenty-five minutes - a record held until the end of the war.

Chapter Twelve
Brindisi

However, as the weeks wore on the role of our Squadron changed. The Normandy landings had taken place and the fury of the German air force was fully occupied on the Western Front. A new role was found. Bomb racks were fitted under each aircraft wing which were to carry 250lb. bombs. These planes became 'intruders', attacking coastal shipping and railways along the coastlines of Yugoslavia and Albania which were ferrying supplies to the enemy. These raids were carried out at pre-dawn and at dusk and for the first eight days were very successful with no losses. A favourite target was the harbour at Corfu which had been attacked three times, but the fourth visit cost us three losses and all the pilots killed.

The following day two more were shot down with one pilot killed and the other bailing out and landing in the sea. Another pilot, making his bombing run, saw the dinghy and radioed back to base reporting the incident. A small Walrus rescue seaplane was stationed at Brindisi for such an emergency. I well remember the Squadron commander rushing into the hangar shouting:
"Pilot in the 'drink'. I want a strong swimmer."
"He's only a few hundred yards from the beach."

One of the lads volunteered, jumped into the Walrus which landed a hundreds yards seaward of the dinghy, swam to the pilot and rescued him while under fire from guns on the land. A valuable pilot was saved who otherwise could have lost his life or been taken prisoner.

During our eight month stay at Brindisi the armies moved north, the city of Bari was recaptured and a Toc H Services Club was opened which operated a canteen and accommodation for about twenty servicemen on leave. It was managed by a Jugoslavian couple, named Belinski, refugees from Zagreb and who depended mainly on volunteers. Whenever I could get a 48 hour pass I would hitch-hike to Bari and give a helping hand.

Toc H Services Club Bari

In due course we left Brindisi and moved up to a major air base at Foggia, almost four hundred miles north - a perilous journey through the mountains where roads were narrow and on many stretches, with very poor surfaces. The first problem was that the Engineering Officer, who was in charge of the move, had no knowledge of handling a convoy, the normal procedure being initially to divide the vehicles into three sections - fast, medium and slow speed.

The slow section moved off first, then the medium and finally, the fast. A separate vehicle with the food and field kitchen would leave sufficiently early so as to set up the feeding point at a predetermined map reference. Another major problem arose due to the fact that the Squadron had never been supplied with any RAF Service Vehicles apart from a fifteen cwt. Bedford truck and two Crossley five tonners. Two Fiat five ton trucks with trailers had been 'pinched' from an Italian transport park and a five hundred gallon tanker on an

International chassis completed the vehicle fleet. One of the Fiats I had converted into a mobile parts store and a 'plug bay' with sufficient space for my bed and personal kit. The other had been claimed by "A" Flight who copied my example. The Commander had already flown his 'kite' bearing the letter "K" up to Foggia.

The orders were for the convoy to move off at 4 am., so that half the journey would be completed in daylight. Apparently the Movement Order wasn't issued until later in the day and it was 4 pm, when it was already dusk, before we got under way. It was mid-October, nights were getting longer and the risk of frost and snow likely. As all vehicles set off together, our speed was that of the slowest one. By 8 pm. we reached the pre-arranged 'food stop' and, hastily grabbing a meal, moved on to a staging post where sleeping accommodation was available.

It was in the early hours of the morning when we arrived absolutely exhausted. Without even undressing we fell onto a bed and into oblivion. Being awakened at 6 am we observed that snow, to the depth of eight inches, had fallen as we slept. A hot meal was served after which we returned to our respective vehicles. Few started at the first attempt, others had flat batteries and what with one problem and another, it was mid-day before we were on the move. Needless to say we didn't arrive at Foggia that day. A second staging post gave us relief for one night and we finally dragged ourselves into the Air Base at 3.30 pm. No wonder we gained the name of Fred Karno's Air Force!

The Rambling party on the cliffs near Ancona

Foggia was indeed a massive airfield occupied by both fighter and bomber squadrons of the RAF and USAAF. For the first time I saw a USAAF Mitchell medium bomber fitted with a 75 mm. field gun mounted in the bomb bay, firing aft. In fact I believe the Americans had every offensive aircraft including Liberators, Flying Fortresses and Bostons on this base. Several Spitfire Squadrons operated from here including some who were part of the Desert Air Force.

Three months later we moved on to Ancona, another short stay before our final move north to Falconara. The war in Europe was now about to end and the Squadron headed south-south-west towards Naples where, in the small village of Gerbini and billeted in a disused macaroni factory, 1435 squadron was disbanded. The next day was VE Day and after a

Imperial Hotel Sorrento used as the RAF rest Camp 1945

parade to the village WW1 War Memorial and joined by the local inhabitants, speeches were made by the Italian mayor and our Commanding Officer. Finally, a farewell address by Flight Lieutenant Lamb, the Adjutant and Welfare Officer since the formation of the Squadron in Malta, was given, tears rolling down his cheeks. I was the only other member of 1435 Squadron who had served during the whole time of its existence. Many of the originals had been posted back to the UK but sadly over thirty of our pilots had been killed in action.

Chapter Thirteen
LIAP & return to UK

I was posted back to Brindisi where 148 Squadron were operating Liberators, flying food supplies to the starving folk of Klagenfort in Austria. In March 1944 the Central Mediterranean Command initiated a scheme for leave back to the UK for long serving personnel in the CMF theatre of war, or those with twelve months yet to serve before becoming 'time ex.' The scheme was designated LIAP and stood for "Leave In Advance of Posting." This scheme permitted leave from the unit for twenty-one days, and I was included. I was in Brindisi at the time and the overland rail and sea journey to Horncastle took seven days each way, so that the actual leave time at home was limited to the remaining seven days. It was, never the less a Godsend.

The 148 Squadron at Brindisi (Wellington's)

After a couple of days in Horncastle, I made my way to Ipswich to meet Joyce for the first time. I had continued correspondence with her since the days in Malta, and what started as friendship had already taken on the overtones of romance.

Our meeting took place in the garden at 45, Park View, Ipswich. Joyce was dressed in a maroon coloured two-piece suit with a silk blouse to match and her hair was beautifully styled. I hugged her and politely kissed her on the cheek. We sat and talked for ages and began to realise that we were well suited to each other and a full romance was bound to follow. The remainder of my leave continued in Ipswich before reluctantly, I returned back to Brindisi.

The next six weeks I was with 148 Squadron which was then disbanded.

My next posting was back to Egypt to Gianaclis some sixty miles west of Alexandria. There I was attached to 23 Squadron flying Mosquitos. As they were equipped with Rolls Merlin engines, the work was almost identical to that of my 'Spitfire days'. There was a heavy flying training programme in progress so that all technical ground staff were kept extremely busy. Servicing the aircraft in desert conditions was too hot

RAF Gianaclis Camp Church

during the middle of the day. Ground staff duty hours were amended as follows:- from 4am to 10am and from 4 pm to dusk.

Whilst at Gianaclis a notice appeared announcing a Moral Leadership Course of three weeks duration and to be held in Jerusalem. I, together with six more colleagues enrolled for the course and a week later left Gianacles en route to Jerusalem. Billeted in St Paul's Church Hall the course consisted of lectures in the morning all of a biblical nature, and visits to the appropriate sites in the afternoon.

We were addressed in the mornings by the leaders of the various Christian religions, including the Bishops of the Greek and Coptic Churches and the Archimandrite of the Armenian Church.

During the last week we went on a three day tour from Jerusalem to Jericho, through Samaria to Nazareth for an overnight stop. On the second day we passed the Sea of Galilee, continued to Capernaum and drove west to Haifa where we had the luxury of a hotel room overnight. The third day saw us following the coast road south to Tel Aviv and then back to Jerusalem. It had occurred to us that the purpose of the course was to condition our minds to a more peaceful existence after having spent six years being engaged in nothing other than killing our fellow man.

On arrival back at Gianaclis and studying the notice board I found my name amongst a list of personnel headed "Postings to the UK".

Two days later a lorry transported us back to Kasfereet Transit Camp, which by this time was packed by "time ex" airmen in the major part of the camp, and recruits recently arrived from the UK in the remainder.

To my disappointment I learned that some of the "time ex" airmen had already been at the camp up to fourteen days but there had been no postings home. The shortage of shipping was one problem. The other was the order in which men were to be repatriated which was based on age and length of service, thus the younger with least service were given higher release numbers. I was Group 36 which meant that 35 groups would be repatriated before it was my turn. This explained why I had been posted first to Brindisi and then to Gianaclis and why, when I did eventually get to the Transit Camp at Kasfereet, a further three weeks elapsed before I was posted on a draft to the UK. Other news we received from airmen now stationed in England really did inflame us. This confirmed that those of the same demobilisation group and based in the UK, had been released some weeks earlier and resulted in a near riot state at the camp.

On the Friday of the third week, extra supplies of beer arrived at the NAAFI stores. By then the numbers of 'time ex' airmen had rocketed since I had arrived. Beer flowed as never before in the crowded NAAFI and by 9 pm the building rocked with bawdy songs in ever increasing volume. The Orderly Officer of the day accompanied by the Orderly Sergeant arrived so as to close the bar but were obstructed from entering. With bellies full of beer, any form of order by the men vanished. Chairs were used to smash windows and, realising the likelihood of further damage, many of us left and returned to our tents. Minutes later the building was on fire and not even the fire engine was allowed to approach.

The building and its contents were completely destroyed. The perpetrators were determined to alert the attention of the Station Commander to the unfair treatment suffered by the men, the same men who had done their duty and helped to win the war over the last six years. In the last three weeks only a handful had been posted home, yet convoys of Army personnel had passed the camp on their way to Alexandria.

At 9 am the following morning all personnel were instructed to assemble outside the HQ building. The flat roof made an ideal rostrum for the Commanding Officer's address. The officer bore the rank of Group Captain. He was dressed in a battered flying jacket heavily decorated, well worn cap and the inevitable 'sweat rag' round his neck. He too, was 'time ex' and like us, sympathetic to the cause. This was not a parade, it was an assembly of very angry men. The shoutings and booings subsided and the instant he took the 'Mike' in his hand deathly silence prevailed.

"Chaps," he started,
"I know your grievance, but wrecking the camp is not the answer."
He had heard that some airmen had been there for five or even six weeks.
"I am coming down to a table below and I shall take

details of those who arrived here three weeks or more ago. Half an hour later I want to see those who have been here two weeks or more. I am now going to give you my word that later in the day I shall inform Command HQ of the disgraceful treatment you have suffered and demand that the entire space on the next troopship bound for the UK will be taken by the men from this camp. In the meantime I expect your peaceful co-operation. I am arranging for a marquee to be erected today to serve as a NAAFI tonight. But let me warn you that if a single man steps out of line I shall have him court-martialled within the hour, and you know what that means. Good morning, one and all, I know you will respond."

It appeared the C.O.'s appeal did not fall on deaf ears for, three days later a convoy of lorries arrived and took a large contingency of us direct to an awaiting troopship in the harbour at Alexandria. The actual numbers were never known but an estimate would have been two to three thousand. We knew neither the name of the ship or our destination until, at sea we learned from one of the members of the crew that we were en route to Toulon. The two thousand mile voyage would take us the best part of three days. That was the least of our concerns. We were, at last, on our way home to the UK and back to 'civvy street'.

Arriving at Toulon we boarded an awaiting train and travelled north but were absolutely amazed at the change of scenery. Being so accustomed to living in the arid desert of Egypt and seeing nothing other than sand and rocky outcrop, to be faced with the brilliance of verdant green fields and lush leafy trees seemed almost unbelievable. But the beauty of the countryside continued, except when passing through towns and villages until, two days later, we reached Calais.

Chapter Fourteen
Joyce & 'civvy' life

We then crossed the Channel to Dover and joined a train taking us to Hednesford, the Demobilisation Centre in Staffordshire. It took but two days to hand in our service kit, have a final medical and to be issued with the proverbial demob suit.

I attended the final pay parade and received cash and vouchers to the value of £225. 0s. 0d., This was made up of back pay, a month's pay in advance and Gratuity pay enabling me to survive for the next twenty-eight days. Finally, I was issued with a railway warrant allowing me free travel to Ipswich. Transport was provided to the nearest station and, having bid farewell to our remaining colleagues, we boarded our respective trains home. I had to travel via London on the old LMS line and take a connection on the LNER line to complete my journey.

I had already decided to make my first port of call to Needham Market where Aunty Ethel lived, as I had not seen her since the tragic affair at Sloothby. Correspondence had been maintained since my days in Malta in which I had learned of her second marriage to a farm foreman. A short bus journey took me to the village and to a small cottage where they lived. It was, indeed, a wonderful experience to meet her again and I immediately recognised how aged she had become in the intervening thirteen years.

Again her cottage was tiny - two rooms up and two down

- with the same primitive facilities as before, and as a teenage step-son occupied the second bedroom, I realised there was nowhere for me to stay. After a cup of tea and a long chat, I left and took the next bus back to Ipswich so as to meet Joyce. There I was warmly welcomed by Joyce's parents who invited me to stay for the time being. Theirs was a three bedroom semi-detached house, built pre-war and, as Joyce's brother Peter was away on National Service, a room was available.

George and Joyce engaged
June 1946

At the time Joyce was nursing geriatric patients in a hospital at Sudbury, some twenty miles away but I soon persuaded her to seek a post in Ipswich. Nurses were still very much in demand in the private sector and she was successful with her first application for a position at Allington House Private Nursing Home. There they dealt with Maternity, Surgical and Medical patients and thus offered a variety of duties much less arduous than her present job. Most of the first three months were spent on night duty but she was able to live at home and we were able to meet for a couple of hours in the evenings as well as at weekends.

Our courtship flourished and on a one week holiday in Hastings we became engaged. My apportioned leave had now almost expired and for me, it was now decision time. What and where was I to seek employment? My post in Horncastle was protected by law and I must therefore see what prospects were there. I also needed to consider the current level of remuneration. Accordingly I re-visited the old firm and met the boss. "How glad I am to see you back, Hows. When are you starting?" he said.

"Just a minute," I replied. "What sort of wages are you now paying?"

"You would be on 1s. 1d an hour", he answered.

"Oh! well I shall have to think about it," I retorted. "It doesn't seem much to me." And I left.

On reflection I soon calculated that with gross earnings of around £5. 5 shillings less tax and National Health contributions, and the cost of rail journeys to Ipswich, it would mean that my courting would be severely curtailed. There must be an alternative, I told myself and as Ipswich was the centre of agricultural engineering, I must see what opportunities I could find there.

On returning from Horncastle I made a few enquiries and to my surprise, discovered there were several retail dealers in the agricultural machinery trade operating within the borough. Both Ford and Ferguson were represented and the local Farmers Co-operative had a machinery department and all were short of experienced mechanics. I visited all three and each offered me employment. I chose the Eastern Counties Farmers as they presented the best prospects. Their foreman was seventy years old and anxious to retire and if I was as good as I had implied, the job would be mine. Thus my first decision was made.

My next thoughts concerned Joyce and our marriage and where we were to live. I was against starting our married life living with in-laws, as I had heard of so many similar arrangements ending in disaster. We decided to hunt for a house that we could buy although our resources were very limited. But here we had luck. As always, its not what you know but who you know that matters, and it so happened that a family friend heard of an old lady living in a bungalow who was anticipating moving into a nursing home. This was confirmed and we rushed round to see her. She took a liking to us and hearing that I had recently left the RAF, she offered the bungalow to us for the sum of £1,150.

"Can you give us forty-eighty hours to arrange the finance," I asked.

"Yes," she replied, and away we went.

Between us we raised the deposit of £250 and took a mortgage for the remaining £900 over 25 years which meant weekly repayments of £1. 2shillings. With an average net income of under £5 I knew things were going to be tight particularly as I insisted that Joyce did not go out to work once we were married, and to this day I am convinced I was right.

This all happened in March 1947 and we then arranged our wedding to take place in May. We were married at St Thomas's Church Ipswich on 9th May. We took immediate possession of the bungalow and were able to redecorate the kitchen, lounge and one bedroom within that period. The next problem was furniture, all of which was in short supply and rationed. Furniture

dockets were issued for this purpose but sufficient only to have a bedroom suite and a kitchen table or a dining suite and a bed. Fortunately Joyce's parents gave us a second-hand bedroom suite and with our dockets and remaining cash we bought a kitchen table and two chairs. A pair of fireside chairs completed our furnishings with which to start our married life.

George and Joyce Married
May 1947

I remember getting a cheap second-hand radio, making two speakers 12 inches square out of plywood and wiring them up to kitchen and bedroom, the radio being placed in the lounge. We managed quite happily and like everyone else, added items of furniture as money became available. We bought nothing 'on the drip' as hire purchase was called.

At the back of the bungalow was a large garden that adjoined a forty acre field. Weeds grew to the height of a four foot hedge and a lot of hard work was necessary before any plantings could take place. I was enthusiastic and determined to convert the wilderness into a productive plot but, needless to say, knew absolutely nothing about gardening. I soon began to learn after all crops failed the first year. It was sandy

Ted Byron Best Man

heath land and needed lots and lots of humus and organic matter and this is how I tackled it. Talking to a farmer one day of my failures, he said:

"You want some muck on there and plenty of it. I bring my empty lorry into the Sugar Beet Factory at Ipswich in October for a load of beet pulp. I will bring you a load of muck if it can be tipped at the bottom of the garden."

"That's brilliant!" I replied. "The bottom fence is in bad shape and needs replacing. I'll make a ten foot section so that it can be lifted out and you can then get the back of the lorry right in."

He did, and each successive year I had a six ton load of well rotted muck to dig in. It worked miracles and I produced enough potatoes to last us most of the year. Vegetables grew in abundance and we always had a choice of at least three roots or brassicas all the year round.

The first year we were married I became foreman at Eastern Counties Farmers with the welcome increase in wages. In February 1948 Joyce presented me with a bouncing baby boy weighing in at 7lbs 2ozs. He was named David George but sadly died with pneumonia when only three weeks old. In June 1949 we had a second son, Robin David, and all went well.

Chapter Fifteen
The inexperienced salesman

This was the year when I started my selling career. I had previously told Charles Westrip, my boss, that I had ambitions in this direction and he held the view that I could make a successful salesman. April 2nd saw me going to work not in overalls but in a new suit, polished shoes, overcoat and trilby hat. In future I was to be addressed not as 'George' but as Mr Hows. It was a strange feeling and not without some trepidation that I started in the sales office studying the literature of the full range of machines and implements together with agricultural hardware and dairy requisites that were sold.

A month later I made my first farm visits – 'cold calling' was the term used. I was to learn a great deal from my first encounter attempting to sell a machine to a complete stranger although I had been tipped off that the farmer was going to buy a new mower and that he preferred to deal with Eastern Counties Farmers. I was told his name and address – Mr Hector Hunneyball, Pond Hall, Wix.

"It's near Dovercourt," said Mr English, the Sales Manager, "and he wants a new mower." Before leaving the office I searched through all the sales literature racks for the various manufacturers producing grass mowers and studied each leaflet until I had memorised the features of each one. I was confident that I was now a competent mower salesman. How wrong I was proved to be! Off I went and found the straggling village of Wix and was directed to Pond Hall by a postman. A

lady answered my knock on the half opened door.
"Good morning," I said, and followed with
 "Mrs Hunneyball?"
"Yes, what do you want?" she snapped
"I've come from Eastern Counties Farmers to see Mr Hunneyball," I answered.
"He's not here."
"He's at the other farm, drilling," she replied somewhat brusquely, but she gave me the necessary directions.

The farm was two fields away from the road with cattle grazing in each so I had two gates to open, pass through and close before reaching the farm. On arrival the only sign of life was in the cowshed where I found the herdsman hosing down.
"Is the boss about?" I enquired.
"Yes, he's drilling in the ten acres". he answered, and pointed me in the right direction.

I crossed the first field which was grassland and it being a windy day lost my hat a couple of times. On reaching the field where drilling was in progress and observing the tractor and drill were at the other end, I waited on the headland for it to arrive. Seeing me he dismounted the drill and approached.
"What do you want?" he asked.
"My name is Hows. I'm from Eastern Counties Farmers," I announced. "Mr English tells me you need a new mower."
"Yes!" he said, "and what can you tell me about mowers?"
I then brought out the first leaflet and repeated all the

features only to get his retort

"I don't want that."

I then brought out the second one,

"This is the Massey Harris power mower," I told him and proceeded to extol its virtues. Again his response was as before. Groping in my pocket I produced a third leaflet saying:

"This is the Bamlett mower, the Rolls Royce of mowers," and I described it in detail. Yet again he replied:

"I don't want that!"

By this time my hat had blown off twice and I was getting desperate. Fumbling in another pocket I brought out yet another leaflet.

"Now sir, this is the mid-mounted mower made by Featherstones. The cutter bar is under the middle of the tractor so you don't have turn your head when it gets blocked."

To my dismay he repeated the comment I had heard so many times before.

"I don't want that!" He then produced a tiny cutting from The Farmers Weekly announcing a new model called the Bamford Major.

"That's what I want. What do you know about that?" he grunted.

"I'm sorry sir but I have no information on that one," I admitted.

"You're a bright salesman, come to me half cocked," he snarled.

However, I knew that he would be at Colchester the following Saturday and I promised I would see him then. This I did. Like many other farmers he always had lunch and a few drinks at The Three Cups Hotel, next door to the Corn Exchange, so I waited until his return before approaching him. He was in a much better mood than when I first encountered him and having given him the details and price of the Bamford Major mower he said:

"You had better get me one," and disappeared.

At last I had secured my first order, but that is not the end of the story. I placed the order with Bamfords stating that it was to be fitted to a Fordson Tractor. Three days later I had a reply asking which spot was it, red spot being low gear ratio, green spot, high gear. I again saw Mr Hunneyball at Colchester Market the next Saturday and enquired:

"Mr Hunneyball, is your Fordson tractor a red or green spot?"

"Buggered if I know. You had better go and look," he replied. Thus my return route home was to be via Wix to ascertain the correct coloured spot painted on the transmission housing.

There was yet another twist to this tale. When the mower was delivered in May I demonstrated it myself to ensure that it performed to the customer's satisfaction. He was delighted with its performance whereupon I produced the invoice.

"Perhaps we can have a square up now then," I said.

"You what?" he snapped.

"Well sir," I explained "I'm told that nothing is sold until it's paid for." We went back to the house and, on entering, he said:

"Would you like a bottle of beer?"

"Thank you sir, I would," I answered thinking, perhaps

he's not such a bad chap after all – one of those whose bark is worse than his bite. I sat and enjoyed his beer and without further ado he fetched his cheque book and paid the bill. I hadn't offended him and he subsequently became one of my best customers.

For the next two years I called on the same farms alternate weeks always on the same day selling mainly small items of hardware and dairy requisites which I could deliver on my next call or, if an order was urgent, would make a special journey after tea. For me, time was never a problem and often I worked until late at night if I felt I could gain a new customer or get another order. My target for the remaining nine months of 1949 was £9,000. But I achieved £14,000. My salary was £350 plus 1% commission excluding the first £3,000 sales. In the second year my target was £20,000 and I achieved £ 28,000.

Three years later our third boy arrived and we named him Christopher George. On the third day he took ill and was unable to retain any food. The doctor attending Joyce prescribed three hourly feeding but vomiting problems continued. The next day the Doctor was called again and to my dismay, insisted that two hourly feeding must be the answer.

Not being happy with the way things were going I stayed home from work. It was a good thing that I did for within an hour of the doctor leaving, the poor child had a convulsion, his colour changing to purple. Joyce was frantic. I rushed next door to ask the neighbour if she would hold the baby while I drove them to the

Borough General Hospital only half a mile away. I was met by the paediatrician who took him away saying; "Wait for about twenty minutes, while I take a look at him."

I was terrified. On his return he called me into his office and lit his pipe. I feared the worst. He then went on to explain the problem. The tube connecting the stomach to the bowel had not dilated, preventing any waste material passing through. There was no medication for this problem and the baby was too weak to contemplate surgery. He said:

"All we can do is concentrate on keeping him alive"

"I've placed him in an incubator with oxygen. It will take time, but we have to let nature take its course. It is a condition known as 'piloric stenosis', and I have come across it once before."

Day by day I visited that incubator only to observe the baby losing more and more weight. Daily I saw the doctor and expressed my concern at the weight loss.

He would reply:

"That is the least of my worries. We are winning, he's still alive."

Twelve more worrying days passed before the doctor asked me to take two ounces of breast milk to the hospital and he would try the first feed. It worked. It wasn't rejected. Two days later the amount was increased to four ounces and that was retained. But I was still worried about the loss of weight. The poor infant looked like a skeleton with no flesh at all on his bum. The next day, seeing the doctor and to my complete surprise he said:

"You can take your son home tomorrow."

"But what about his weight?" I protested.

"No need to worry about that now. He's as strong as an ox. His Mum will soon put weight on him" he commented as he headed for his office and another pipe of 'baccy'. I prayed he was right, Chris was 6lbs 4ozs at birth and had gone down to a pathetic 4lbs 2ozs, but he soon put on weight as forecast. And that is how our second son survived.

Chapter Sixteen
The McBain inventions

George with Mc Bain Pea Cutter. A. Mc Bain Inventor,
C. Westrip and F. Moss E.C.F.

In the next year I became interested in something entirely different. One of my customers, a Mr Andrew McBain was developing a machine for cutting and windrowing peas in a single operation. Having a special interest in anything akin to engineering I spent lots of time helping in the design as I could visualise great potential in the marketing of such a machine. The only ones available at the time were of a complex design and very expensive. This machine made use of the three-point linkage and power take-off on a Ferguson tractor and could be produced at a fraction of the cost of any competitors. My interest was to obtain the manufacturing rights and launch it on to the UK market.

Many discussions ensued with the Executive Committee of ECF but I pressed my case hard. In the end I was called into the General Manager's office and given the go ahead on three conditions – (1) I could build up to ten machines. (2) all must be sold. (3) a profit must be made by the year end. I accepted the challenge and with a lot of help from Andrew McBain, met all conditions. In fact twenty-five machines were built and sold as I discovered when purchasing raw materials and components than few suppliers were not interested in orders for ten.

I demonstrated machines in every pea growing area in the country making sure that at least one was sold in each. The following year the figures rose to one hundred although I was given authority to produce only fifty. The machine was first exhibited at the 1953 Royal Show at Windsor and was seen by all the UK dealers as well as visitors from overseas. This resulted in orders being placed by Dutch and French importers for a minimum of twenty-five machines each. Every year sales increased as did the number of our employees including a Production Manager. I handled all retail and wholesale business appointing dealers in the area where volume sales warranted. In the export market I dealt only with importers who, in turn, set up their own dealer network.

As the success of the McBain Pea cutter continued 'Mac', as Andrew McBain was now known to us all, turned his inventive mind to a more ambitious project. Would it be possible to invent a one operation harvester to handle the green pea crop? The current system of cutting, loading, transporting the crop to the nearest static viner where the peas were despatched to the factory and the residue returned to the farm was laborious, time consuming and expensive. If cereals could be harvested this way, why not green peas?

The problems to overcome were many due to the tender characteristics of the crop. Green peas are of a very delicate nature and were easily damaged on the one hand, whilst a certain amount of brute force was required to expel the peas from the pod on the other. Added to this was the fact that during the vining operation the laceration of the haulm released lots of sap creating a sticky mixture of peas, pods, leaf and stalk that had to be separated. The static viners were huge machines with no limitation of dimension or weight. For a machine to be mobile, severe limits were placed on both.

'Mac' started with the chassis of a scrap Oliver combine, to which was fitted a cutting and elevating mechanism. This part worked well. The vining mechanism, the heart of the machine, consisting of three slowly rotating cylinders above concave grills effected the vining process. Theoretically, the peas fell through the grills but, inevitably some were carried through with the lacerated haulm. Development over the next three years improved separation to an acceptable level and prototype machines were field tested in Norfolk, Lincolnshire and Lancashire before being shipped to Australia and the USA.

In 1957, I entered the McBain Pea Viner in the Royal

E.C.F. Stand Royal Show Bristol 1958 Mc Bain Pea Viner
wins the Burke Trophy

Agricultural Society's Silver Medal Competition for the Best New Machine or Appliance of the Year. The Judges spent a great deal of time and expertise scrutinising the machine but insisted that they wished to witness field performance before reaching their decision. A demonstration was arranged and met with their full approval. Accordingly, in 1958, a second application was made resulting in the award, not only of a Silver Medal, but also the Roland Burke Trophy. This was the highest award possible and of great prestige value.

We were on the threshold of a spectacular success as none of the competitor's world wide had progressed on these lines. Imagine my response to the reaction of the Eastern Counties Farmers executives who did a complete "U" turn.

Press Cutting outlining the Australian Mc Bain Pea Viner Field Trials

In spite of this profitable business, many farmer members of the Association objected to the manufacturing nature of our work and at the next meeting of the Executive Committee, voted to close my operation down. When Charles Westrip informed me, I was furious at the decision. I was offered the job of setting up a new department to handle the sales of all second hand tractors, combines and machines bought in by the seven machinery branches. But, knowing that many had been traded in at unrealistic prices there would be no chance of that department making a profit, this was not for me. I made my mind up there and then I would seek employment elsewhere. I felt disgusted at the shabby way that I had been treated. There was no future for me in this organisation.

I was by now well known throughout the wholesale and retail trade in the UK and I now made use of this. I let it be known that I was seeking a change of employer and within a week was offered the choice of three appointments – two in the retail trade and one in wholesale. It was while I was considering these that I

Cyril Hack Managing
Director Tractamotors Ltd

received a letter marked Private and Confidential. It was from Cyril Hack. We had met at Stoneleigh Abbey at a Ferguson seminar for Dealer Principals staged six years earlier. In the evenings a few of us met in the bar and swapped a few beers discussing the topics raised during the day. The letter was informing me of Cyril's promotion to the Board of Directors of Tractamotors and the vacancy for a General Manager. If interested I was to ring him and have a chat. This I did and it resulted in Joyce and I being invited to Melton Mowbray for the weekend. We met at the new Tractamotors premises in Scalford Road which were as yet unoccupied.

Tractamotors new premises 1959

Chapter Seventeen
Tractamotors

Detailed discussions took place and we adjourned to the George Hotel where Cyril had booked a room for us. In the evening a dinner had been arranged at The Bell Hotel that was attended by Cyril and his wife Kathleen, Peter and Mary Drewry, the Bank Manager (whose name escapes me) and his wife, and Joyce and me. Peter Drewry, who was a local influential farmer, sat next to me and turned the conversation to almost every aspect of farming and I was conscious of the fact that he was endeavouring to see how extensive my knowledge really was.

The evening was a happy and successful one for Joyce and me after which we returned to the George Hotel. On the Sunday morning we had a walk round the town and after lunch, Cyril drove us through the surrounding area. Then we were invited to his lovely house named Cerignola (that name rang a bell) in Burton Lazars for tea. The ladies adjourned to the lounge and Cyril and I to a small room called a 'snug'. Here we discussed a final proposal for my employment which I accepted. Fortunately, although it meant that Joyce had to leave her beloved Suffolk, she agreed instantly. We were both very happy to move to Melton and in retrospect I have to say that I had the feeling that Cyril was a man I could trust, and that together we two could make Tractamotors an even more successful enterprise in future years.

I joined the Company in April 1959 with no

management formal training but convinced that I would succeed in my new post. Cyril obviously had confidence in me and I was determined not to disappoint him. I had just to turn up the learning curve. At this time we had no sales representatives and I undertook these duties until new appointments were made. Monday, being market day in Stamford, I made this my first priority. It was an important market and was attended by most of the Rutland farmers together with many others from surrounding areas. I approached the cattle pens where the auctioneer was selling fat cattle and noted who was there.

Enquiring from a bystander, I obtained a few names and when the auction ended, I saw several farmers adjourn to a nearby bar which was part of the George Hotel complex. Giving them time to settle in, I went in and bought a drink. Identifying Mr Frank Gilman, I approached him and introduced myself mentioning Cyril Hack by name. It wasn't long before I had the names of all those present who included Charles Gilman (Frank's brother), "Barty" Aldwinkle, Alfie Roberts, Ted Makey, Henry Griffin, "Pinkle" Andrews and others whose names escape me. I had at least made a start, sort of broken the ice; not easy in an area completely unknown to me.

I left Stamford without scratching an order in my book. So, studying the local map, I motored towards Oakham on the road to Exton. Following a sign to Cuckoo Farm I found the farmer at work around the stack yard. I introduced myself and saw that he was using Ford tractors. He didn't seem to know Cyril

Hack and always traded in Stamford. I let him talk all about his farm and the family and at last he revealed that his next purchase would be a crop sprayer. That was just what I wanted. Coming from East Anglia, where every farmer had one, I was familiar with the whole topic of crop spraying including choice of chemicals and rates of application. I convinced him that he could do no better than place an order with me for an Allman Model 60 sprayer that very day to ensure that it would be on the farm well before he needed to use it. The order was placed and I scratched it on the first blank page of my book. The name was Wakerling, known as Bob. It made my day. Not only had I got my order, but Tractamotors had gained a new customer.

During the next few months no new representatives were appointed, Massey Ferguson were suffering with a continuous fall in market share due to the disastrous reputation of the Grey and Gold model of the MF35 tractor. Numerous failures of the clutch and final transmission together with engine starting problems gained this model a bad name. By May 1960 many modifications were introduced including a new engine, the Perkins 3A152 diesel.

A Mr Larry Pomeroy appeared on the scene. Having just returned from Australia where similar problems had been experienced and working under contract with MF, the market share had been raised to the acceptable level. His commission was to make a similar achievement in the UK. Within 24 hours of arriving in Coventry on a Friday afternoon he arranged to meet every distributor at pre-arranged venues the following day. His programme was made possible by flying to each of the venues by helicopter.

At the distributor meetings a six week campaign was announced when we were all given a sales target. To the representative who achieved the most sales of tractors there would be a prize of either a new refrigerator, a television or a cheque for £70.

I was still the only person selling for Tractamotors so that the pressure was really on. But I had a bit of luck. Rupert Froggatt, a local farmer and motor trader had just bought Ragdale Hall Estate and ordered six tractors and the day before I had sold two to Henry Hibbitt who farmed at Empingham. None had been registered at the factory. This gave me a good start. I checked our records to ascertain every owner of a Ferguson tractor in Leicestershire and Rutland. Day by day I made the calls and presented the case for replacing the 'little Grey Ferguson' with a brand new FE35, with much improved power, a six speed gearbox and ground speed power take-off. My technique was successful and almost every day, I came home with an order for one or more tractors. At the end of the campaign my total was thirty-six.

Six weeks later a sales meeting was held at The Bull Hotel, Peterborough to announce the campaign results. The room had been equipped with floodlights, spotlights and a series of charts which unrolled during the report. Larry Pomeroy presided and seemed in high spirits. After declaring the campaign to be a tremendous success and thanking all who had participated, he proceeded to announce the winners. The man who took third prize was from Weatherheads of Cambridgeshire. The spot light focussed on him, a fanfare blasted out and he approached the rostrum to receive his award. The second Prize was awarded to a man from Rogers of Northamptonshire.

Sandy Thorbourn receives his retirement car

By now I was feeling very disappointed. With my sales volume I had expected to be either second or third but suddenly, the spot light picked me out and I was declared the winner. I went up and was congratulated and presented with a television set but then realised that I could buy the same model from Garners, the local TV dealers in Melton Mowbray where I would get 10% discount! I went back to Larry to say that I preferred to take the cheque and he readily agreed. Having returned home I made my purchase at Garners at a net cost of £57 showing me a profit of £13.

The increased level of sales continued and must have reached the ears of Sandy Thorburn who prior to my arrival, was a sales representative for Tractamotors but had left to join Cottesmore Motor Company selling Ford motor cars. This move had not proved successful and he was anxious to rejoin us. After due consideration he was re-engaged and allocated the area east of Melton and including the whole of Rutland as he was already well known to the influential farmers there.

A few weeks later I was in Oakham market and observed a middle-aged man standing on his own looking particularly miserable. I hadn't met him before so decided to make myself known to him. I said. "Hello! I'm George Hows, and I am with Tractamotors"
He replied. "Oh yes, I'm Jack Hughes and I'm with a small fuel oil company. It's owned by a contractor at Thorpe Arnold called Vernon Lovegrove and things haven't been going very well. In fact he is closing it down and I've got the job of getting the money in."
I enquired, "what are you going to do then?"
He replied. "I'd like to get back into the farm machinery business again." "I used to be with Sharman and Ladburys and before that I was with Garners."

I responded. "That's interesting, Tractamotors have a vacancy for a sales rep. at the moment. If, when you're finished with this job and you're not fixed up, give me a ring."

Ten days later he knocked at my office door and with Cyril's agreement was engaged and given the area west of Melton to operate.

Chapter Eighteen
Grain storage & drying

We now had the salesmen needed. Both were experienced and required little training, so it allowed me to turn my attention to a new venture, grain drying and storage.

My plan was to sell such a plant to an influential farmer near to Melton. I had already sold a tractor to Mr Francis Johnson, at Moscow Farm, Great Dalby and in conversation, had gleaned the information that now he had a combine, he was thinking of storing his grain instead of having to sell it 'off the combine'. This would enable him to sell his crop at a much higher price in the following Spring. After a long discussion regarding the various crops he was growing and their acreages, it was possible for me to make a recommendation as to how many and what size the silos should be. I then raised the question of the grain dryer most suitable and the necessity to have a good cleaner/grader.

Next day came my offer to prepare working drawings for a plant, suitable for all his requirements, fully costed and without obligation. He agreed to these proposals and I proceeded to arrange dates for delivery and erection of the plant. There now followed daily visits to the site so as to monitor progress. Thus my plan succeeded and I had sold the first grain drying plant in the East Midlands. This was to be but the beginning of a very successful enterprise. Growth was bound to follow. A manager was appointed to take control of the newly formed department.

The Hows family at Julie's Christening 1962

The next major event to occur in our family was at 8.30 am., on Tuesday, March 13th 1962 when Joyce presented me with a daughter weighing 6lbs. 13ozs., I was ecstatic. It was cloud nine for me, and I don't think I've been down to earth since. That very same afternoon I had a 'phone call from Loughborough informing me that Robin had been rushed to the Hospital in Baxtergate for an emergency operation to remove his appendix. It never rains unless it pours, or so they say. After visiting Joyce in St Mary's hospital I raced over to Loughborough to see Robin and this was to be my routine for then next eight days. During this emergency, 'Buddy' the wife of a friend of mine Ken Gore-Graham kindly undertook to look after Christopher after school and to provide us both with a hot evening meal. It really is wonderful how there always seems to be someone to come to your aid when the occasion arises. I was so grateful.

In 1962 Massey Ferguson together with Butler Engineering formed a new company to market the Massey Butler system known as Stor-and-Dry. Imported from the USA it consisted of circular silos arranged in an arc with the grain being circulated by the use of grain augers, the complete system being out

Massey-Butler Silo installation Kennilworth 1962

of doors thus not requiring a building. Some of the silos were fitted with drying floors as necessary and others purely for storage. This was a relatively low capital cost system when compared with current traditional ones available in the UK. Sadly the warehouse arrangements at Witnash, where the imported components were held, proved chaotic. Despatches were short shipped or wrong components sent. Although for two seasons we were obliged to live with these problems, we developed a substantial volume of business, but at Witnash their problems increased resulting in Butler Engineering being forced to close down.

A competitor, Spiroflight, operating from Potton in Bedfordshire, immediately jumped in to take advantage of the situation. I applied for the franchise and was appointed a main distributor. The silo business thrived and the terms were exceptionally favourable enabling us to make unusually good profits. But then calamity . . . . Spiroflight went into receivership. Yet here was another opportunity. We negotiated a hard deal with the liquidator and bought the entire stock, and at our offer price. This enabled us to wholesale as well as retail all silos, electric motors and augers until none were left. A very profitable operation.

In the meantime, the manager of the Grain Drying department of Tractamotors diversified into a series of projects including specialist farm buildings for grain "on the floor" storage and drying, potato storage, milking parlours, silage silos and automatic feeding systems. Each of these required special knowledge and skills that the manager did not have and that, together with insufficient "on the site" supervision led to serious problems. Building work was unsatisfactory and, in some cases had to be done a second time. Installations didn't function to the customer's satisfaction and payments were withheld.

The result was that profits made in the grain silo business were more than amortised by these blunders. The decision was taken that it would be in the best interest of the Company if the department was closed down.

By now Tractamotors was established as Volvo car distributors, and a petrol station had been opened on the forecourt. Both enterprises were contributing to the Company's success.

I was now Managing Director and Cyril, although still remaining active in the business had gone to live in Kent. Far from fading out of the picture, he continued to guide me and offered help and advice when required. A superb relationship had, over the years, developed

116

between us. It was more than a partnership. We had become great friends and after forty three years, the situation is still the same.

The next development in the Company's history occurred when Cripps, the local Land Rover dealer closed down. As soon as the news broke I rang the Marketing Director of Land Rover seeking the dealership. Already he had received six other applications so I had to work hard to press my case. I sought an interview the following day and this was agreed. The next day I made the first of many visits to the Land Rover factory at Solihull. Three days later the Sales and Service Directors came to view our premises and facilities and were sufficiently impressed to confirm Tractamotors as their new distributors.

They then informed me that it was their intention to terminate Victor Woods in Oakham and Chandlers at Grantham dealerships so as to have a more viable one at Melton Mowbray. Their future planning was to have Land Rover Centres at strategic locations thus having a much wider customer base so as to justify the greater investment in the modern premises required.

We recruited the necessary staff and soon discovered there were more orders on our books than the supply from the factory could meet. The situation in the Service Department was quite different. We were overwhelmed with work, most of which was warranty, inherited from the previous dealer. The latest model was the Land Rover 90 which had lots of problems, mainly water leaks in the bodywork. Fortunately warranty rates were the same as our retail, so that we still retained our required profit level.

It was at this stage that we realised we had outgrown the three-quarter acre site at Scalford Road and, as both Volvo and Land Rover were demanding a 'solo' site, new locations needed to be found. But I was not to be part of these developments.

Chapter Nineteen
Joyce's illness

For several years my wife, Joyce, had a series of health problems. Having made a full recovery from open heart surgery in 1984, there followed a lengthy period of psychological illness, and finally she became a victim of breast cancer.

For the next twelve months Joyce enjoyed much better health although I was aware that this was a period of remission from cancer. How long or short this time would be we knew not. Most of that year was spent either on holiday or going out for the day. We had a wonderful holiday in Pembrokeshire with glorious sunny weather. We had a memorable day in Lydstep Bay that was surrounded by high chalk cliffs. These had been terraced to accommodate residential caravans on hard standings and connected to all services. Car parking was just above beach level and ideal for the limited numbers of day visitors. One could sit in the car and enjoy watching families on the beach and the sailing and water ski-ing in the bay. Having taken lunch in the restaurant nearby, we returned to the car as clouds gathered and a storm threatened.

A number of yachts were moored on the west side of the bay and a rescue team was always available should any craft get into trouble either at sea or when moored. Gale force winds blew in the heavy storm that followed and one of the yachts broke away from its mooring. High waves and wind carried the yacht across the bay heading for the rocks on the opposite side. The rescue team had difficulty starting the engine on their boat and that delayed their launching.

Once under way they raced after the threatened yacht but alas, it had already reached the rocks and was being badly battered. The crew, realising that it was too perilous to proceed, had no option but to abandon their attempt and return to their own mooring. In the next twenty minutes we witnessed the total destruction of what had been a lovely sea-worthy yacht, but the storm soon cleared and we enjoyed the rest of the day in brilliant sunshine.

During our fourteen days in this area we explored the beautiful South Wales coast line as far as St David's. There we visited the cathedral, had tea and drove back to Saundersfoot where we were staying. Two other days were spent at Lydstep Bay which we enjoyed so much.

On our return home we revisited our favourite spot on the east coast, Sutton-on-Sea, on many occasions. I had been to Sutton since my boyhood days at Horncastle when boys could hire a bicycle from 'Nunky' Brown (so named as he had a hare lip) for twopence a day. Sutton has changed little over the years and remains quiet and peaceful, unlike Mablethorpe or Skegness.

It was on our return from Sutton that, on reaching Coningsby Joyce complained of a serious headache. We stopped at a restaurant for a cup of tea and I went to the chemist across the road and bought some

Paracetamol tablets. Joyce took a couple with her tea and we continued on our journey back to Melton Mowbray. A restless night ensued and I called the doctor next morning. He prescribed other medication which seemed to be ineffective but he told us that twenty-four hours must pass to give it time to work. The following morning Joyce's condition suddenly deteriorated and the doctor was called yet again. He immediately ordered her to hospital but as no beds were available in any of the Leicester hospitals, five hours elapsed before one became available. An ambulance took Joyce to the Leicester General Hospital.

Subsequently Joyce was taken to the Royal Infirmary for a brain scan and rushed to the Queen's Medical Centre at Nottingham. At 7 pm the consultant neurologist telephoned me with the dreadful news that Joyce had developed a brain tumour that was inoperable, and that she had but a few hours to live. The family gathered at the hospital and she passed away at 1.15 am. next morning. We were devastated and after the funeral I grieved for two to three weeks and then came to terms with the fact that life for me would never be the same. We had shared forty-two years together but that had now ended. I had to reshape my future life style.

Within a year of my retiring age, I resigned my appointment as Chairman of Tractamotors. Though heart broken at this decision, I had no choice but to care of my dear wife with whom I had shared so many precious years.

The future of the Company was to be in the hands of a younger man, Cyril's son-in-law Peter O'Connor, who was a keen young man and who Cyril considered could be trained as a potential successor.

At this point I have to say that Cyril was most kind and considerate concerning my personal problems. No one could have been more so. I was to remain a Director of the Company and retain the "perks" that I had previously enjoyed. What was even more valuable to me, he continued to be my best life-long friend.

Chapter Twenty
Agricultural Shows

In the previous part of this story there is a dearth of information surrounding my personal and family life so that I will now try to fill in some of the gaps. In my infancy I was brought up to be a Christian and, having been confirmed at the age of twelve, I have always endeavoured to be faithful to its principles. In times of trouble I have always found help and comfort, good fortune, or something, which seems to have led to individuals who have befriended me. The social organisations with whom I have been and am still associated, have greatly influenced the direction of my life.

Over the passing years I had come to know that many men who had shown me kindness or helped me were freemasons. These included G. W. Waymouth, Headmaster at Cagthorpe School, Charles Westrip, my manager at Eastern Counties Farmers, Roland Achurch, my employer, and other members of the local community. All were held with high esteem and it became my intention to enquire further about my becoming a member.

On one occasion when discussing a work programme with Charles Westrip, I questioned him on the subject. He informed me that, whilst he would be happy to propose me into his Lodge, there was a five year waiting list. But four years later I moved to Melton Mowbray. A full year elapsed before I realised that Cyril was a member of The Rutland Lodge and I was able to convey my ambition to him. The waiting list here was three years and, in due course, he proposed me for membership and I joined in February 1963.

I became Worshipful Master in 1977. As the years went by I joined other Lodges and Christian Orders but had insufficient time to fully support them. Now I had the time to further my interest in this important work at the same time making many new friends. The aims of freemasonry in no way conflicted with my religious beliefs and, in many respects, reinforced them. This explains why I still support the movement to the extent that I do.

Rotary played a similar role in my life though was less demanding of time. The needs of the voluntary sector caught my attention as it was the time when the Melton Volunteer Bureau was desperate for drivers. These were needed to transport patients to hospitals in Leicester and Nottingham. This work appealed to me and I joined the Bureau. I have since found it both rewarding and enjoyable, at the same time feeling that, at long last, I could repay society who had given me so much over the years.

My sons were ten and seven years old respectively when we moved to Melton. Robin started at the Brownlow School and at the age of eleven, was accepted as a boarder at the Endowed Schools, Burton Walks, Loughborough. Christopher went first to the Sarson School then on to the King Edward VII Grammar School.

In June 1964 I was mown down in Burton Street, Melton when two cars racing each other approached the bridge. I suffered a broken leg and multiple lacerations to the face. The ambulance took me to the War Memorial Hospital where Doctor Manson patched me up. The following morning I was taken to The Royal Infirmary in Leicester and finished with my leg encased in plaster from toe to hip. It was the worst time of the year to be out of action as we had four grain stores in the process of being erected on various sites around the area. However, Cyril arranged for a driver to chauffeur me around.

The Royal Show was held at Kenilworth during the first week in July and it was arranged for me to assist on the Massey Butler Stand. This was vital to me as I was bound to collect some orders for silo installations for the following year. John Hall, the Marketing Director of Massey Butler, realising my predicament, sent a car to collect me and booked me in at the Welcome Hotel at Stratford-on-Avon for the duration of the Show in order that I could attend. From my point

E C F Stand Suffolk Show Ipswich 1949

of view it was a very rewarding operation as I logged six excellent enquiries which, subsequently, matured into firm orders.

Ever since the war I have been involved in Agricultural Shows, not only at county level but also on the national and European scene. My first experience was at the Suffolk Show held at Christchurch Park, Ipswich in 1949. I was then the workshop foreman and was responsible for arranging the machines and implements on the Eastern Counties Farmers Show Stand. My boss had instructed me to attend the show on both days so as to answer the technical questions which might arise when farmer customers were considering making a purchase. A huge marquee complex had been erected adjacent to the display area in order that hospitality could be dispensed. Food and drinks, including alcohol, were generously served by the office staff acting as waitresses.

The stand was busy all day but, needless to say, most of the farmers visiting were mainly interested in the food and drink. There was no control of those entering, and the system was very much abused. In some instances a farmer would bring the whole family in on the first day and all his workers with their families on the second. So often it would be the ones who spent the least with the Association during the year who consumed the most on Show days.

The trade stands closed at 5 pm and I was busy covering a grain cleaner down for the night when an old chap wandered on to the stand. With knees out of his trousers and elbows out of a jacket which had seen better days, he lingered around a combine harvester parked next to where I was working.

"Hey boy!" he said, pointing to the end of the cutter bar. "What's that bit do?"

"That's the divider," I replied.

"What's it do?" He questioned?

"That lifts the crop and the knife cuts a clean edge avoiding blockages," I answered. He walked round the machine and continued asking questions for about ten minutes. I tried to continue covering the grain cleaner and noted that he was now studying a baler parked next to the combine. Again he posed a series of elementary questions and I provided satisfactory answers, but by this time I was wishing that he would make himself scarce. All the staff were in the Staff Tent supping their ale and I was anxious to join them. This scruffy fellow – a farm worker I presumed, - was wasting my time. But he persisted. I was just about to leave him when he said:

"Just a minute, young fella. What does one of them combines cost?"

"Thirteen hundred and twenty-five pounds," I replied.

"And what's the price of that baler?"

"Six hundred and ninety pounds," I told him as he pulled an empty cigarette packet out of his jacket pocket and a pencil stub from his waistcoat. He scribbled the figures down and enquired:

"I sink you'll knock a bit off?" Taken completely by surprise and realising the total cost was £2,015, I said "If a farmer ordered one and paid cash in advance, yes, I could."

"Well get on with it," he replied, fumbling in the inside

pocket of his jacket and pulling out a cheque book. "Alright," I said. "Make it £2,000."

"Here you are then. You'd better make it out," he said. I wrote the cheque out and he endorsed it with a scrawl. It was then I realised that he couldn't write very well. He told me his details.

"My name's Hadingham, Ted Hadingham, and I farm at a place called Metfield, near Halesworth. It's forty miles or more from here. You'd never find it. The only way you can get to it is off the runway of the airfield at Metfield. I've got about twelve hundred acres and my boys have a bit of land next to me."

"How interesting," I remarked. "I'm in the Service Department and one day soon I'll come to see you. What sort of tractors do you have?"

"All sorts. I buy 'em at farm sales when I see one I like. I've got three or four Fords, an old Allis Chalmers and a big Caterpillar."

"How nice to meet you, Mr Hadingham," I said, "I really must go now. I'll be in touch. Goodnight!"

It was now almost seven o'clock and I made for the tent where all the drinking was taking place.

"Where on earth have you been, Hows?" asked Westrip.

"Doing a bit of business, Guvn'r," I announced.

"Reckon I've done in the last hour more than anyone here has done all day. I've sold a combine and a baler to a new customer, and there's the cheque."

"Well done!" said he. "You'd better get yourself a beer to celebrate"

This taught me a few lessons. Sequel - I did find Ted

Hadingham's farm, albeit with great difficulty, and he became one of my best customers. He had four sons for each of whom he had bought five hundred acre farms. He bought collectively for them all. I took orders for binder and baler twine, a ton at a time; and six, forty - gallon barrels of lubricating oil once a year. After I demonstrated an Allis Chalmers tractor with an under slung steerage hoe, he ordered one for himself, costing £725., and one each for the boys. This was the best deal I ever did. I didn't give a penny discount.

Chapter Twenty One
San Salvador

The show seasons commenced in May and continued through the summer to September. Eastern Counties Farmers attended those in East Anglia starting with the Suffolk and followed by Essex and Norfolk. These were all two day shows. In addition, several smaller one day shows were staged at Tendring Hundred in north east Essex and Aylsham in Norfolk. Later on when I was involved with the McBain Pea cutter, I attended most other County Shows where dealers were exhibiting them.

These were the days when most shows were mobile, moving around the various parts of their respective counties. Sites varied as to suitability, often in parkland, so that one might find a mature tree in the centre of the stand site. Or it might be on a slope covered with mole hills. Such were the hazards of setting out the stand in weather that was not always favourable. This is why I enjoyed showmanship. On arrival at the Eastern Counties Farmers stand site, I first made myself known to the occupiers of neighbouring stands offering any assistance if it was needed. Later I had no hesitation in shouting for help.

The system worked well and a great rapport built up between us. It was always important to develop a good relationship with the local Police responsible for security on the Showground. On arrival I always informed them of our stand location informing them that they would be welcome and that bottled refreshment was available in addition to tea or coffee. This worked wonders as few stand holders offered hospitality before the Show opened. In 1953 we took a Stand at the Royal Show for the first time, exhibiting the McBain Pea cutter. It was held at Windsor that year and was favoured with scorching hot weather.

The regulations of the Royal Agricultural Society were very strict here and a band of Machinery Stewards constantly patrolled the stands to enforce them. Grass had to be closely mown, uniform size signs erected to standard heights and all stands enclosed within a given perimeter of posts and chains of similar height. All work had to be completed by the Saturday night for final inspection by the Stewards on the Sunday prior to the Show opening the following day. All this information had been detailed in a Rule Book issued when the Stand space was booked so that there was no excuse for any infringement. I always ensured that all our signs, fences and posts met with the strict requirements before we left our depot. However for those who never read the Rule Book, problems occurred and in extreme cases the culprits would be excluded from showing. But it was the part of my work which I thoroughly enjoyed and in most cases found very rewarding. Sales were made at the shows in those days and I seldom returned home with a blank order book.

In 1975 hay and straw was packed into bales of medium size by the traditional range of balers in use at the time. These were not easily loaded, transported or stacked and a change of the system was called for. I felt that the time had come to investigate other systems appearing elsewhere, particularly in the USA.

Jeremy Coleclough and George
San Salvador 1975

Tractamotors was already selling the Danuser post hole digger marketed by Opico, based at Spalding, Lincs. The Chairman was Jeremy Coleclough, son of the boss of Rotary Hoes, both of whom I had known for many years. Jeremy was thinking on the same lines at the time and considering a visit to the VI International Fair to be held in San Salvador in the early part of 1975.

At the Smithfield Show at Earls Court I visited the Opico Stand and met Jimmy Oppenheimer, the President of Opico whom I had also known for several years.

"Hi George!" Jimmy greeted me, and after a couple of drinks, enquired:

"Why don't you come with us to San Salvador?"

"Why on earth should I want to go there?" I replied.

"We're going to the Fair," he explained.

"You would enjoy it. I'd love you to come. Be my guest."

I couldn't take his suggestion seriously as I realised that he had imbibed, probably heavily, from the whisky bottle, and dismissed the idea. However, on my return to

Jimmy Oppenheimer
Founder of the
Opico Corporation

the office, a letter from Jimmy awaited me. He was serious. He did want me to go to San Salvador as his guest. Furthermore, the phone rang and it was Jeremy Coleclough the Managing Director of Opico Limited. He'd received a copy of the letter that I had just read and was instructed to persuade me to accept. I had now to seriously consider the invitation particularly as it would give me the opportunity to evaluate the choice of American built big balers.

I discussed the matter with my Managing Director, Cyril Hack, who readily agreed that I should go. Accordingly I accepted, and the necessary plans were made. Jeremy confirmed travel arrangements and on the appointed day accompanied me to Heathrow Airport to board a Jumbo Jet bound for Washington. From there we took an internal flight to Atlanta. On our arrival he rang Jimmy who lived in Mobile, Alabama and we were persuaded to take a further flight there immediately.

As there was an hourly service from Atlanta we took the next plane and on arrival at Mobile, hired a car and drove direct to Jimmy's home. There we were greeted by a welcoming party consisting of most of the VIPs of Mobile. Dinner followed and by then I was very travel weary. With a seven hour time change, it was now 3 am

GMT but only 8pm local time. But Jimmy insisted on extending his hospitality by taking us to a late night roadhouse for music and dancing. An hour later I said that I was all in and took a cab back to Jimmy's and retired.

The next day more of Jimmy's reps arrived and in the evening were entertained to dinner at the Bienville Club, an exclusive establishment on the 14th floor overlooking the central railway station, harbour and the State Throughway. During the meal I had a spectacular view of a departing freight train of one hundred and fifty-nine trucks hauled by four diesel-electric locomotives, two pulling and two more at the rear. A delightful meal was served - five star service - enjoyed by all.

Another late night but this seemed to be the American way of life. After a supposedly relaxing weekend in Mobile, we drove about a hundred miles south to New Orleans and made a bee-line to the Pontchartrain Hotel to assuage the thirst brought on by the near tropical weather. There I was introduced to Raymond in the cocktail bar who specialised in his own menu of cocktails. We partook of his gin fizz, combining $1\frac{1}{2}$ ozs. of Gin, $\frac{3}{4}$ozs. sweet and sour, 2ozs. dairy cream, 1 teaspoonful of sugar, 1 egg white, $\frac{1}{4}$oz. orange juice and soda, blended together with crushed ice and served in a Collins glass.

Having drunk a second gin fizz I began to realise how potent a drink it was. One more and I would have been flat on my back. We then adjourned to the Hilton Hotel and stayed the night. The following day we attended a major agricultural exhibition at the Marriott Hotel. This, I suppose was the equivalent of our Smithfield Show with this difference, no machines were exhibited and all presentations were on video. Each exhibitor took a hotel suite that served for both business and hospitality, a very practical system and much less tiring for the visitors.

During our short stay in New Orleans I walked the length of Bourbon Street, the heart of American jazz and called in a number of jazz halls.

The final part of our journey was a flight to San Salvador, calling at Belize in Honduras en route. On landing at San Salvador, we could see over two hundred light planes parked on the perimeters. This indicated to me both wealth and opulence but on the way to the Camino Reale Hotel, we saw nothing but abject poverty. Natives survived in cardboard boxes or under rusty sheets of corrugated iron. Dressed in rags, they were begging for food - a distressing sight indeed.

The hotel was the venue of the Opico Conference and all arrangements were made by Inman Ellis, Senior Vice President of the Company. The format of the conference was in typical American style. We started early at 8am with presentations of new products by the suppliers and ended at 12 noon. A mobile bar was then wheeled into the room and drinking commenced followed by a light lunch. The afternoon was then free until dinner at 7 pm., and partying until as late as we liked. The very hot weather did seem to be the cause of so much thirst.

One of the delegates was Eddie Valenzuela, the local Massey Ferguson Distributor, in whom I had a particular interest. We were able to compare our respective businesses which were of similar size, the only difference being that we had diversified into non-MF machines whilst he had a successful air-conditioning franchise. He invited me to his dealership and I was able to spend a whole day there when every aspect was openly discussed including the problem of warranties. These were a major problem to him, as to reach the situation where payments to MF were being withheld. Most of the tractors supplied were manufactured in Argentina and these were the troublesome ones. A few originated from the Coventry plant and it was possible to make comparisons. The time had come when Eddie decided to accept tractors only from the UK plant.

Eddie had his own private aircraft as did most business folk in San Salvador. I was thinking that the visit was complete when he rang his mechanic at the airport ordering him to prepare the plane for flight. We drove down to the airport and he gave me a most interesting trip at low level over El Salvador, Nicaragua, Guatemala and Costa Rica. I was able to observe various crops including cereals, cotton and coffee plantations on the mountains. This was a most exciting and informative experience that I appreciated.

During the week I was introduced to Henry Danuser, maker of the post hole diggers, and whose factory was in Fulton, Missouri. Jeremy had known him for years and had stayed at Henry's before so that it was not surprising for us to be invited to spend the weekend there. We accepted although it meant changing our plans and foregoing our holiday in Miami.

There was no problem in changing our flight programme so that after the conference we flew back to New Orleans, then took a second internal flight to Dallas and on to Kansas City. There we hired a car and drove over three hundred miles west to Wichita, Oklahoma, to visit the Hesston factory. Here we were able to see round balers and other harvesting equipment being produced. Leaving our car at the airport we flew back to Kansas, hired another car and drove to Fulton Missouri and the Danuser ranch.

I was reminded that it was at Fulton where Winston Churchill gave his 'Iron Curtain' speech at the time when the bombed Church of St Mary's, Aldermonbury since rebuilt in the grounds of the University, was re-dedicated. This church was to be the memorial to Winston Churchill and the ceremony was attended by President Truman and the Archbishop of York, the Rt. Revd. Cyril Garbett. During my stay, Violet, Henry's wife, took me on a tour of the church, to the crypt in which was an exhibition of the life of Winston Churchill. Henry, on his part showed me a film of the dedication ceremony.

We toured the factory under the guidance of Henry's son who was now taking over the business from his father. Early one morning I wandered around what I would call the farmyard but was much more extensive, and saw lots of vintage machines and implements. So ended our stay at this delightful ranch home, we drove back to Cleveland, Ohio, and then took a flight to Kennedy Airport, New York and finally boarded the plane for our homeward flight.

No reference to this expedition would be complete without a tribute to Jimmy Oppenheimer. Once described to me as a wandering Jew, he certainly had a unique modus operandi. With an extensive knowledge of agriculture, both ancient and modern, he travelled worldwide to identify the local needs for every piece of equipment known to him. He would then contact the appropriate manufacturer in the USA, obtain quotations and set up a supply line. As the business developed in each country or continent, a representative was appointed. What to me was incredible was that, having built up a worldwide business over twenty-five years, not once had he called his representatives together for a sales meeting. This was the very first one and he included a request for the managers of all his suppliers to participate. I was in the unique position of being the only guest.

Chapter Twenty Two
Retirement & Voluntary Work

The outcome of this visit was the decision to market the Hesston Round Baler, well proven and popular in North America. Although a completely new concept, there appeared a distinct advantage in the round bale being more weatherproof than a square one. Furthermore, straw in round bales was easily unrolled in covered cattle yards, and the only extra handling equipment required was a single spike attachment to either front or rear loader. I was convinced that it was a winner and would suit many farmers in the Leicestershire and Rutland areas.

The first season, six were sold and gave reliable service. The following year I gave Jeremy the biggest order of any dealer at the time - for fourteen machines - and I was able to negotiate extremely attractive terms. By now other manufacturers had copied the round baler principle but being well established in the market, this affected us very little. Thus my trip to the States with Jeremy proved a good investment.

Returning now to family and coming to terms with the loss of my dear wife Joyce. Some hard thinking had to be done. However I was fortunate in having my daughter and her family living nearby from whom I received tremendous support. But Julie and family had their own lives to live and I was determined not to become a burden to them.

I was sixty six and enjoying good health. Life was now

good to me in spite of an unhappy start. Society in the form of folks throughout the years had done much for me. Now was the time that, perhaps, I should do something for others. There was an abundance of opportunities in the volunteer sector. I opted for the Melton Volunteer Bureau that was in desperate need of drivers to convey patients to the hospitals in Leicester and Nottingham for treatment not available locally. This I have continued to do for the last sixteen years and I still find the work most rewarding. One meets so many interesting people of similar age and who are so anxious to chat about their experiences during and since World War II. Most patients are very appreciative of this service but occasionally one gets the odd moaner. I choose to remember those in the former group. Here is an example.

Two maiden sisters well into their eighties were to be taken by me to the Ophthalmic Unit at the Leicester Royal Infirmary for an annual check up. Both were retired professional ladies – one a nursing sister and the other a school headmistress. One was called Lisa and the other Mary. It was a pouring wet day and with umbrella raised I knocked on the door which was opened by Mary. Lisa was halfway down the stairs and ignoring me, shouted:
"Not the red one, Lisa, the blue one."

Lisa retraced her steps and reappeared carrying a huge hold-all together with an umbrella. Mary, armed with similar items, escorted her sister under my umbrella to the car to take their seats, one beside me and the other in the back. Off we went and before we were out of

Melton Mowbray both were busy eating sandwiches taken out of their respective bags. Then each produced a bottle of orange juice and proceeded to take a drink. On reaching the hospital I enquired:

"Do you know the way to the Ophthalmic section?"

"Yes!" said Mary. "We come here every year."

It was still pouring with rain and I escorted them to the hospital entrance informing them that having parked the car, I would rejoin them. Parking at the Royal Infirmary is difficult and I had to queue before securing a space: it was probably twenty minutes before I was able to commence my search for them. Arriving at the waiting area I found them, and again their bags were open and they were munching away to their hearts content before taking another swig of the orange juice.

In the process of the afternoon, three different sections were visited each with its waiting area allowing my patients further opportunities to partake of their refreshments. It was nigh on five o'clock when their visits were completed and I gave them clear instructions where to sit just inside the main entrance whilst I collected the car. Returning to the main entrance, I was told by an attendant:

"You can't park there!"

"I'm collecting a couple of elderly ladies who are just inside the main entrance," I replied.

"Alright then, but make it quick," he retorted.

I entered the hospital only to find they were not there, so I went back to the Ophthalmic department to find them. Eventually I did so and escorted them back to the car and saw them seated with seat belts employed.

Torrential rain continued and the traffic was at its worst. I endeavoured to find my way out of the city centre only to discover that Belgrave Road was closed and all traffic had to seek alternative routes. An hour later I found a way to the A46 and headed for Thrussington. In the meantime I realised my passengers were busy searching their purses as I heard the tinkling of coins, and by the time we reached Rotherby Top, a handful of change was being handed to me.

"What are you doing?" I enquired,

"I don't take tips. I do this job for the joy of it. Now you put that money away."

An argument went on as to which one had given this coin and which had given that, but at last all seemed to be resolved. Mary, sitting next to me, now started rummaging in her handbag and produced one of those old fashioned conical sweet bags (which must have been there for years) and insisted that I should have two of the sweets stuck inside. After all that I didn't have the heart to refuse.

Chapter Twenty Three
Rome Tour

A further interest of mine has always been travel. During my time in the RAF I was fortunate to visit many places abroad including Capetown and Durban on my way to Egypt and thence to Luxor, Thebes and the Valley of the Kings. Whilst serving in Italy I went to Naples, Vesuvius and Pompeii. Shortly after the fall of Rome I was due for fourteen days leave and was anxious to visit this important city. It so happened that a fitter on our squadron named Jimmy McCabe had a brother who was a priest at the San Quattro Monastery in Rome, and I was asked to contact him.

My leave was given and I hitch-hiked from Brindisi to Rome by courtesy of a daily courier plane operated by the Americans. The other passengers on board were either commissioned officers or nurses destined for the Allied Headquarters in Rome. On arrival at the airport, some six miles from the city centre, the other passengers left the plane and were met by several staff cars and driven off. I was the last to leave only to find a remaining staff car driven by a lance corporal who enquired of me whether Brigadier So-and-so was on board. Having told him that I was the last one on this flight, I asked if he was from the Allied Headquarters and if so, could I beg a lift. He agreed and I was driven in style, which made a change for me. He stopped at the huge white marble memorial to King Emanuel II, known to the troops as 'the ice cake'. There I stood with a kitbag full of dry rations scrounged from the squadron cookhouse before leaving, having heard that food was desperately short in Rome.

Being declared an 'open city' no servicemen were allowed to stay overnight in Rome and a transit camp had been established a couple of miles away. I stood for a few minutes cogitating on my next move when a young Italian boy approached and said;
"You want hotel, Johnny? I take you to Swiss hotel. You have food, yes? You follow me."
He took me to the hotel; I handed my kitbag over, was issued with a civvy suit and charged 2,000 lire (approx. £20.00) for a fourteen day stay. Most of the guests were Americans and the others, Naval Officers. All had contributed food of which the Yanks had plenty and we fed like 'turkey stags'.

I made my way to the San Quattro Monastery, enquired after Father Kevin McCabe, and was admitted and welcomed by the Abbot. I explained the reason for my visit and he informed me that Father Kevin was at prayers until 4 pm. In the meantime he invited me to join him in a glass of wine. I accepted and he chatted about life in Rome during the war and then made a suggestion saying;
"If you wish to see the most important places including the Vatican, I will release Father Kevin each day from 2 pm., to guide you. He has been with us for the last fourteen years and knows Rome well."

Such a generous offer I could but accept. Shortly after 4 pm., Father Kevin appeared. He was Irish, stood over six feet tall with 'ginger' hair and was delighted to have news of his brother, Jimmy. He took me on my first tour of the Forum, the ancient market place of Rome, and left me at 7 pm. On subsequent days I had conducted tours of St Peter's Basilica which included climbing inside the dome to the spherical room at the top capable of holding nineteen people. From the ground it appeared to be a gold ball. An 'audience' with the Pope Pius X11, previously known as Cardinal Pacelli, was arranged for me together with nine other visitors. He spoke in English and seemed to have a good command of the language.

On another occasion we spent the whole day in the Vatican including the Sistine Chapel to see the spectacular work of Michelangelo. As Kevin was known to the priest on the door of the Treasury that was kept locked, we were allowed in. The amount of gold and silver, beautifully embroidered robes and vestments was breathtaking. Most had been donated over many years by Catholic communities throughout the world. This was indeed an unforgettable sight that few members of the public ever see.

Another interesting visit was to the Palace via Venezio where Hitler and Mussolini held their meetings. All windows were covered with steel shutters for protection. A tiny balcony overlooked a small square below which might have held fifty people yet propaganda films superimposed St Peter's Square which held a quarter of a million.

The Catacombs of Caracalla were of great interest as these were where the early Christians hid from the Romans before the time of the Emperor Constantine.

Being conducted personally round all the sights of Rome was so informative and enjoyable that they are ever etched in my memory.

Chapter Twenty Four
Madrid Conference

Those experiences evoked in me a continuous interest in sightseeing wherever and whenever an opportunity arose.

My thirst for travel seems never to be assuaged be it local, further afield or abroad; be it buildings such as castles, palaces, churches or cathedrals. Many are impressive and all are treasures. Often, one is attracted by the travel brochures that now bombard us and away we go to distant lands, this at the expense of missing the treasures on our door steps. For example, I have only recently visited the Leicester Pumping Station to see the magnificent Beam Engines, over one hundred years old, a marvel of the engineering of our predecessors. Another treasure is the two thousand horse power steam engine at Wigan, in Lancashire, which drove all the machinery of a cotton mill. Yet another is the host of wind and water wheels used to grind wheat into flour or pump water to great heights or distances. So one can go on.

Fortune has always come my way when it comes to travel, first during World War II, later in my business career and finally as an adjunct to retirement. Whilst exhibiting at the Paris Agricultural Show during the early fifties I enjoyed visits to Notre Dame Cathedral, Montmartre, the Louvre as well as theatreland including the Moulin Rouge, Folies Bergeres and the Lido.

Business visits to overseas exhibitions in Utrecht Holland, Strasbourg, Germany and Verona Italy gave me the opportunity to see many interesting sights, some to be revisited in future years.

Many trips abroad immediately post war were with my great friends Don Rose and his wife, newsagents in Ipswich. Don, having served as National President of the British Newsagents Federation, always attended the European Conference that was hosted in turn by the various countries. Joyce and I were invited as guests of Don, our only expense being the air fares as the host country bore all other costs. The first one we attended was held in Madrid, Spain being the host country. Apart from the opening and closing sessions, we saw little of the conference but a great deal of the many trade stands.

On arrival we met other UK delegates and were entertained to lunch by the Foreign Office and Tourist Department at a five star hotel in a fashionable part of the city. Tables were circular and for ten diners. A five course meal was served with an ample choice of wines, and brandy accompanied the coffee. Amongst the guests at our table was an Italian lady seemingly on her own. We noted that she was always the first to be served but as she spoke no English we were unable to ascertain her identity, until the waiter arrived with the brandy, the bottle covered with a linen napkin and she made a scene. Tearing the napkin from the bottle she swore in Italian at the waiter, who promptly disappeared and returned with a bottle of Grand Marnier VSOP. We thus deduced that the lady in question was of some importance and it was soon made known that she was none other than the wife of the Conference President, Signor Gabrionelli, an Italian.

This was a most unusual situation as it was the custom for the host country to provide the President. Gabrionelli was apparently a very powerful man in the newsagents world in Italy and controlled all outlets in Rome, Naples and other major cities. It was even rumoured that he was a Maffia man and had demanded that he should preside over the Conference so as to enforce his will on any contentious issues.

Later that day we were entertained by the Italian TV and Radio organisation who tried to out-do the sumptuous luncheon. Again the wine flowed generously followed by brandy. Each day we were taken on sightseeing tours including excellent lunches and dinners. One day there was an invitation to a Champagne Party and lunch at the Lord Mayor's Parlour. For almost two hours the champagne flowed and caviar was served before another feast of a lunch was served.

At 4 pm. we were taken to a brand new printing plant that was to be commissioned in our presence but not until we had been hosted with more champagne and caviar.

Having inspected the modern plant and seen the first papers printed we were then whisked off to the RAC motor race-track for yet another drinks session but this time a traditional cocktail party. But this was too much even for the most hardy of our party. Never before have I seen so many folk refuse free booze. There must have been about two hundred people in the party who later sat down for dinner.

As always Gabrionelli presided and fourteen waiters paraded huge barons of beef around the dining room before the main course was served. What a day! It was a marathon. Perhaps it was good that the following day the Conference was closed and we returned home.

Chapter Twenty Five
Peru

Several years later in 1992 I enjoyed a visit to Peru, which was of particular interest. I was accompanied by John Roper who I first met at the Rotary Club of Melton Mowbray. He too was a widower and had recently lost his daughter with M.S. John was a great traveller. He discussed with me a visit to Machu Pichu and I found the idea something of a challenge particularly as he was over ninety years old and would need some help. In fact, it was really company that he wanted.

We flew from Heathrow to Amsterdam as there were no direct flights from UK airports. There we boarded a Boeing 747 on the seventeen hour flight to Lima with one re-fuelling stop at Aruba, a tiny island in the Netherland Antilles. When we arrived in Lima we stayed at the Sheraton Hotel. From there we visited less affluent parts of the city where we saw something of the poverty and neglect of what had been a beautiful modern city. All public buildings were heavily guarded with tanks and armed soldiers. Police were all carrying revolvers, not in their holsters, but held ready for instant action.

A single day in Lima was enough for us before we left on our flight to Arequipa. This meant flying over the snow capped mountains of the Andes and from Arequipa we flew on to Juliaca Airport. Here we boarded a mini bus and passing mud houses observed donkeys, sheep, cattle and llamas grazing on land showing no sign of greenery. No wonder they were all so scraggy or that the cattle produced little more than a pint of milk a day. The climb through the mountains was picturesque but there was little sign of life apart from nomadic tribes wandering with their mixed herds.

Finally we arrived at Puno, 12,562 feet above sea level and felt the affects of altitude sickness. John had his 'Ventolin' inhaler to hand and I was glad to share it with him. Checking in at the Hotel Isla Esteves we were instructed to rest for at least six hours so as to assimilate the rare atmosphere of the altitude. In the evening a taxi took us to town where we found the Hostal Sillustani restaurant serving a typical Amerindian meal. I was too tired to enjoy a meal and, together with the shortage of breath, ate very little. Freshly crushed oranges produced an excellent drink and I was glad to retire as soon as we returned to the hotel. Sleep was not easy and after two hours I went out on to the balcony and watched the bird life on the water. They included gulls, cormorants, coots, herons and Andean pigeons. Several small fishing boats were also active at this early hour.

We were of course, on the banks of Lake Titicaca, a lake with an area of 3,500 square miles. This is the highest lake in the world at over 12,000ft. After breakfast we boarded a small boat that took us to the floating island that was made entirely of reeds with forty-six houses and a school. Everything constructed of reeds which grew in abundance on this part of the lake. A complete community lived there and there was even a pig farm. Boats, used for fishing and collecting reeds, were made of the same material and, when they rotted away, new ones were made. The livelihood of the population seemed to depend more on the tourists than on the pig farm though many of the women made and sold small toys of straw. There was no NHS but twice a week a boat paid a visit to Puno where a doctor was available together with a pharmacy.

Our next journey was of one hundred and twenty miles which included thirty stops to the Peruvian City of Cuzco and scheduled to take anything between ten and sixteen hours. The first stop was at Juliaca where trucks were shunted on and off our train and took nearly two hours to do so. Crossing a flat plain that once was part of Lake Titicaca we skirted a mountain range passing snow-capped peaks some twenty thousand feet high through magnificent scenery. The valleys through which we passed varied tremendously - some arid and dry whilst others were well watered and able to support herds of sheep, cattle, llamas and alpacas. A few were cultivated producing crops of potatoes, leeks, broad beans and dahlias.

As we lost daylight all curtains on the coaches were drawn as raiding gangs were known to be operating in the area approaching Cuzco. These gangs of terrorists, known locally as "The Shining Path" operated throughout the country and attacked under cover of darkness. They could destroy a single village and all its inhabitants in one attack. Or they moved into part of a town or city and massacred people and destroyed property. Hence the need for so much security provided by both the military and police forces.

We only experienced one of the attacks they made on the city of Lima. Awakened one night we heard the rattle of machine gun fire in the distance and the reply from the security forces. The battle lasted almost an hour and then all went quiet once again. Next morning the local bush telegraph spread the news that sixteen innocent people were killed in this skirmish. Tourism to Peru was seriously affected due to the political situation and the operations of "The Shining Path" terrorists.

Cuzco, with a population approaching forty thousand, had but two industries - a brewery and a Coca-Cola bottling plant. We visited the main city square, Plaza de Armous, the Cathedral, La Componia and St Christopher churches and the San Domingo monastery. All buildings were the work of the Incas. Many stones hewn from a quarry a mile away were between two and three hundred tons in weight and perfectly shaped to form walls without the use of mortar.

We then went on to Saqsaywaman - in Quecha, - the local language; or translated into English, 'sexy woman'. This was an enormous fortress with three walls built to protect Cuzco from attack. Sufficient land was enclosed to house and feed the twenty thousand inhabitants of the day. When the Spanish Conquistidors had arrived in the 16th century, stones were removed from Saqsaywaman to build the cathedral and twenty-six churches, yet the fortress walls were still ten to fifteen feet high indicating the enormity of the original structure. Our tour continued to Puca Pucara, another fortress in the River Urubamba valley built to protect Cuzco. That night was spent at the Savoy Hotel, Cuzco.

Next morning we boarded an early train bound for Machu Pichu. The narrow gauge railway took a zigzag route up the mountain face as we left Cuzco until, at about three thousand feet, the track was traditional, winding its way along the mountain edge through absolutely breath-taking scenery. Three hours later we arrived in Machu Pichu. There we boarded a Dodge coach which threaded its way up the mountain side round twenty-four hairpin bends at a frightening speed until, at 7,740 feet we reached the site of the 'City of the Last Kings' or the 'Inca City'. This was discovered in 1911 by the American explorer Hiram Bingham. Believed to have been built in the 15th century, Machu Pichu had a population of twenty thousand and was the last refuge of the Incas. The site had become a jungle having been uninhabited for centuries but subsequent excavations revealed agricultural and industrial areas, houses for ordinary people and Royal apartments for kings and princesses.

The mountain side was methodically terraced to facilitate the growing of crops, grain being stored in huge stone silos. Prisons and torture chambers have been unearthed giving some indication of the awful punishments including garrotting used in those days. Machu Pichu was the reason for our visit to Peru and we enjoyed every minute. We returned on the train to

Cuzco where we stayed the night. Next morning our plane bound for Lima left at 7 am., and arrived at 9.20 am., where we stayed at the Sheraton Hotel.

A couple of days there gave me the opportunity to visit the local Rotary Club. They met in a huge mansion heavily guarded by tanks, armed soldiers and sentries. Having proved my identity and been given a 'Visitonte' badge, I was escorted to the room where the meeting was being held. Ten out of the eighteen members sat round a table each with their glass of wine. They were engaged in noisy conversation but when I made my appearance, silence reigned. I then realised I had a problem. I spoke no Spanish and no one seemed to know the English language. Such difficulties had to be overcome and with a smattering of Italian, a few gestures and a pencil and paper, we were able to converse. The only difficulty arose when I was endeavouring to describe the difference between a pork pie and a Stilton cheese. It was great fun and I enjoyed the visit immensely. That was Rotary at its best.

The following day we took another early flight from Lima to Iquitos - a wonderful flight across the Andes and along the River Amazon to The Explorama Camp there to stay for three days. The Amazon is over four thousand miles long and at this point, over a mile wide. We stayed in wooden cabins on the river banks. Here the humidity was saturating and with the temperature over ninety degrees Fahrenheit it was far from comfortable. Mosquitoes abounded as did lots of other airborne flies and insects but all buildings were netted so as to keep them out. A two inch gap at the bottom of our cabin door was forgotten, permitting the ingress of an iguana. Having persuaded the offender to leave, I used a spare blanket to fill the gap. I hung up my clothes hoping they would dry during the night only to find they were dripping wet next morning.

A native guide who spoke good English, having spent a year in London, conducted our tours for the next three days. Various tribes were visited one of which was the Yagua found on a small island ten miles down river from Explorama. Native women were topless and men and boys wore only a loin cloth. Poison darts were used to kill birds perched high in the trees and harpoons for catching fish. A small clearing made in the rain forest gave sufficient space for their village. There was an abundance of bird life including parrots, parakeets, black starlings and toucans.

One morning we went up river and branched off into a small tributary to try our hand at fishing for piranha. The guide issued each of us with a long stick with a short line and hook already baited. We parked under overhanging trees and fished, and fished, and fished but only the guide made a catch, albeit but four inches long. At least we had seen a real piranha.

Another day we visited a modern village built by UNICEF on a two mile clearance alongside the Amazon, named Indiana Village. Fifteen hundred natives lived there in modern houses comprising, a school, hospital, dentistry and shops. All were connected with concrete roads which were kept spotlessly clean.

On yet another small island we found a wonderful selection of butterflies, and on one river trip we observed hosts of wild birds. They included; white collared hawks, humming birds, Dielda moth bird, Wada Jacana, Dona Cubis, Red Shoulder Macaw, Orio Bird, Ringed Kingfisher, Amazon Kingfisher, Road Si Hawk, Yellow Head Caracara and Turkey Vultures. What a wealth of birds to observe in their natural habitat!

All too soon our tour ended. We returned to Iquitos, took a plane back to Lima and the following day flew back to the UK via Amsterdam. Thus ended a fantastic experience ever to be treasured and remembered.

Chapter Twenty Six
Euro Tour

Bim Bellamy

We planned an entirely different travel holiday in 1996. I teamed up with Bim Bellamy, a friend of many years standing and who had lost his wife two years after I lost Joyce. Bim had earlier shown interest in touring the Czech Republic, and this resulted in my planning a tour across Europe to Prague returning through Germany, Holland and Belgium. Bim agreed to do all the driving whilst I navigated, booked hotels and acted as treasurer.

The Channel Tunnel was opened the previous year and this was our route from the M25 to the E40 motorway in France. Our overnight stay was in Brugge where we spent the next day sightseeing. Sights included the Gothic City Hall, the Bell Tower and the Brugge Carillon with its 47 bells - total weight 27 tons. Then on to the Courts of Justice, the prison and the Tanners House.

Leaving Brugge on the E40 we journeyed on to Ghent, Brussels, Liege, Aachen, Cologne and Koblenz - some 180 miles, with an overnight stay at Hotel Trier Hof. On departing from Koblenz on the E44 we drove to

Greifenstein to visit an underground bell museum, the castle and the church where choristers were rehearsing. Listening to the magnificent singing, we were tempted to tarry, but time was flying and we had planned to reach Stuttgart for dinner. Making our way along the A52/8 we drove past Mainz, Mannheim and Karlsruhe and reached the outskirts of Stuttgart. We continued on to Sindelfingen where we stayed at the Novotel Hotel. After freshening up we dined on tomato soup, salmon steak with Hollandaise sauce, washed down with a glass of lager and finished with ice cream and coffee. It was a wonderful day but very tiring and we went to bed at 10.30 pm.

Next morning we had an early breakfast and drove to the Mercedes Plant where we had booked a conducted tour, due to commence at 9.30 am. The Reception Area was luxurious with a beautiful white Mercedes 280 Drop Head Coupe as the centre piece. We joined a party of about forty and after a film show, were conducted to a 'motorised train' with five carriages, each with twenty seats, which drove us on a spiral roadway to the top of the factory. There we could view the computerised Components Store and, as we returned, the robotised Body Shell Section and then finally to the assembly line of the completed vehicles. Air-conditioned and spotlessly clean, this was an example of excellent working conditions for all and was most impressive. The tour ended at exactly 10.30am and the next party were already assembled. This Plant was producing sixteen hundred cars a day and some nine hundred tons of sheet metal were used for the body shells.

Later we drove into Stuttgart and attended a meeting of the local Rotary Club. At 2.45 pm we left Stuttgart on the E52/8 motorway and headed for Augsburg, scene of the first war-time daylight air raid by Lancaster Bombers. I had been involved in servicing one of those that flew from Coningsby and whose target was a factory producing engines for the German submarines. That raid was disastrous as out of the twelve aircraft sent, only five returned. One of those crashed on landing killing the entire crew. Very little damage was sustained by the German factory. As a result, the idea of daylight bombing was abandoned by the RAF. For myself, I shall ever remember the sight of the crash and the many explosions that followed rendering any attempt at rescue far too dangerous for ambulance and fire-fighters.

We travelled to Munich, the scene of the 1938 crisis meeting of our then Prime Minister, Neville Chamberlain with Hitler, which was later heavily bombed by the RAF. The city had been rebuilt with modern looking buildings, unlike its pre-war appearance.

We carried on towards Salzburg and on to Berchtesgaden where we stayed the night. Our intention next morning was to visit The Eagle's Nest, Hitler's mountain hide-out, but it was closed. Driving back to Salzburg we checked in at The Weisses Kreuz Hotel for a planned two-night stay. Next morning we visited the cathedral, visible from our room, studied the modern bronze doors in relief and designed on the theory of Faith, Hope and Charity. The West Front was flanked by two huge towers in Untersburg marble and was dominated by figures of Christ, Moses and Elijah. Entering the cathedral we were impressed by its size and the richness of the marble stucco and paintings. In the evening we were able to go to the theatre and we enjoyed a wonderful presentation of 'The Sound of Music'. What better way of spending an evening in Salzburg?

The following morning Bim was able to obtain a replacement credit card at the local American Express office for the one he had lost and we then searched for the restaurant where the Rotary Club met. This found, we learned that unlike most Clubs who meet at midday, this one met in the evening. Our plans were amended and we then visited Mozart's birth-place, now a museum. All the captions on the displays were in German but a couple who spoke English fluently kindly interpreted them for us. In the evening we attended the Rotary Meeting and were warmly welcomed.

Our next venue was Vienna and we took the scenic route known as the Salzkammergut. We passed the Mondsee lake at the foot of the Drachenwald cliffs, a picturesque spot, and on to Bad Ischl. Then on to Ebensee at the foot of Lake Traunsee and Traunskirken to the Landhotel where I had once stayed for a holiday. There I was able to renew my acquaintance with the proprietor, Wolfgang Groller.

The journey continued through magnificent mountain scenery skirting wooded foothills cut by wild ravines until we joined the motorway leading to Vienna. We stayed at the Hotel Artis on the outskirts of the city which had an underground car park and the trams running by. Bim was a very good driver and so far his performance had been immaculate, that is until I observed he was driving down the tram lines. Furthermore a police car was parked in a side street as we flashed by. Perhaps they were having a nap as they did not pursue us and eventually we were back on to the correct part of the road.

The following three days in Vienna gave us time to visit the Spanish Riding School, home of the famous Lipazzaner horses, St Stephen's Cathedral, the Baroque style church of St Peter and the Schonburg Palace, exquisite gardens and Museum of Carriages. We even found time to visit The Imperial Palace of Schwarzenburg which is now an exclusive hotel. Here the Rotary Club of Vienna meet with its members of high society. After the meal a learned address was given by the Chief Rabbi entitled 'The Jews and Communism in Today's World.'

We left Vienna and continued on the E5 motorway to Bratislava, Trencin and Zilina staying at the Hotel Slovakia. Weather deteriorated during the night and it rained heavily most of the day. With low cloud, driving was most unpleasant. Having planned to spend a couple of days up in the Tatra Mountains seven thousand feet above sea level, conditions caused us to make a change and drive a further one hundred and fifty miles to Bryno. Here we searched for a hotel, not easy at 8 pm on a Saturday night in a foreign city.

Turning left at a set of traffic lights a Police Car appeared with lights flashing and sirens wailing and ordered us to stop. We had, apparently, done something wrong, but what? Not allowed to turn left at the traffic lights frantically indicated the 'coppers' who spoke no English and us no Czech. However, after offering our apology by signs and showing our need for sleep, they very kindly escorted us to the Continental Hotel where we checked in.

By morning the weather had improved and under a cloudless sky we continued on the E50 motorway through Letovice and Vtavy to Litomysil where we had lunch. Our next stop was at Hradec Kralove and then Podebrady about ten miles from Prague, our ultimate goal. The Golfi Hotel had been recommended to us and there we stayed three nights.

Prague is a lovely city with the River Ultava passing through the centre and crossed by the ornate King Charles II Bridge. We saw the Changing of the Guard outside the Palace as we made our way to St Vitus Cathedral and Wenceslas Square and the Town Hall to see the Astronomical Clock, which dates back to 1410 AD. The other sites we visited included The Great Horse Market, the House of the White Unicorn and the Auto Club where we attended a meeting of The Rotary Club of Prague.

On leaving Prague we made for Dresden, but on reaching Terezin, had a most disturbing experience. This was obviously a military town with a huge central square surrounded on three sides by barrack blocks, six storeys high. Walking round we observed a black, cast iron plaque announcing that three hundred Jews were imprisoned here before their execution. On the fourth side of the Square was the headquarters of the infamous German SS who, in the war, had administered the notorious concentration camps. On leaving the town and turning an 'S' bend we stared directly into one of the dreaded extermination ovens. We were both horrified at this sight which accounted for the uncanny atmosphere we had previously experienced as we walked around.

We proceeded to the German / Czech border just beyond Teplice. There were long queues of lorries awaiting inspection by the frontier guards but cars passed through without delay. Our next stop was Dresden, the city completely destroyed during the War, but now rebuilt. A single night stay here was sufficient and next morning we joined the E40 motorway leading to Chemnitz and Jena. Here we took a minor road to Erfurt to visit the cathedral, where fifteen hundred years ago a twelve ton bell was installed. My particular interest in this bell was that a replica was cast by Taylors of Loughborough and became the memorial to "The Siege of Malta" in World War II.

We continued our journey through Weimar and skirting the town of Naumbers we had a narrow escape from a serious road accident. A car suddenly emerged from a petrol station on our right and a collision seemed inevitable. Bim braked hard and our car shuddered to a halt with the clatter of metal and vibration from the rear. We were certain that impact had occurred. Imagine our relief when we discovered that the two cars were but an inch apart and the sounds were made by a beer can trapped between the rear wheel and the wheel arch, and the vibration was due to the cobbled road surface. The offending driver apologised and explained that the setting sun had blinded him. The next town was Weissenfiels, and as Berlin was still seventy miles away we stayed at the Jagerof Hotel for the night. It in no way appeared to be a hotel as it was built as a castle, then converted to a nunnery.

Next morning we continued through Mersenburg and Halle and spotting a roadside barbeque and bar, we stopped for refreshments. The proprietor was playing a piano accordion and as soon as he recognised our nationality, he struck up with 'The White Cliffs of Dover' followed by 'The Quartermaster's Stores.' Other guests were German and were very friendly and joined in a sing-song of popular tunes.

Resuming our trip we arrived at Potsdam, some twenty miles south of Berlin. Here we stayed at the Ramada Hotel Berlin Tetlow and visited Berlin the following day. Scrounging free seats on a tourist coach we arrived at 10 am., then took a city tour of four hours duration with an English speaking courier. Touring the city centre we passed the zoo, a ruined church now known as 'The Rotten Tooth,' a reminder of the horror of aerial bombardment, the new shopping centre named Euro-Centre, the Japanese and Italian embassies and the German and French cathedrals.

We were shown the Jewish ghetto and a remaining section of the Berlin Wall, then travelled down the Unter Den Linden to the Brandenburg Gate. A short stop there gave us time to buy souvenirs and take photos. Returning to the City we passed through the Tiergarten, a parkland area and the World War II memorial. Thus our tour ended and we then had a meal before meeting the coach for the return journey back to the hotel.

The following morning we set out on the E30 Potsdam - Berlin Highway but traffic was very heavy and roadworks prevented overtaking. We were stuck for miles behind an ancient diesel truck discharging clouds of fumes travelling at a snail's pace. Passing by Brandeburg and heading for Magdeburg there was still no escape from the awful fumes, but to break our journey we had lunch at the next Service Area. Thus refreshed and rid of those fumes we continued through Osnabruck and Ibbenburen where we found a smart Austrian style hotel and made it our overnight stop. At dinner the food was excellent and the room first class.

Next morning we headed for Arnhem and visited the War Museum and cemetery at Oosterbeek. We both had memories of the disaster of the Battle of Arnhem when, out of the ten thousand paratroopers sent, fourteen hundred were killed and six thousand taken prisoner. Watching the film it was heartbreaking to see the wholesale slaughter of so many young men some not yet twenty five years of age. With sombre hearts we toured the rows of graves, checking each one, searching for names of two men known to Bim. So many were engraven with the words 'An Unknown Soldier' and the names we sought were not found.

Leaving Arnhem in a heavy thunderstorm we drove to Nijmegen scene of another bitter battle in an Allied attempt to cross the Rhine. Then on to Antwerp and Ostend where we stayed the night at the Soll Cress Hotel in the small village of Anjy. Our eyes had been caught by a display of vintage farm implements and of chickens, ducks, geese and a couple of ostrich and we later learned that the hotel was hitherto a farmhouse.

A bantam cock herald the morn from 5 am., which motivated us to make an early rise and with breakfast at 8 am., we started the final stage of our homeward journey. Heading towards Dunkirk, traffic was reduced to a crawl due to roadworks and we had to follow a diversion. Making our stop in Dunkirk and visiting the Saint Eloi's Church where the main body of it was on the right hand side of the road and it's tower some forty feet away on the left, indicated its original size. After lunch we made our way to Calais and calling in the Duty Free area continued through the Channel Tunnel. Thirty five minutes later we arrived at Folkstone and drove back to Melton Mowbray by the shortest possible route. So ended a memorable expedition across Europe in which we had visited seven countries, ten capital cities, six cathedrals and driven three thousand seven hundred and fifty miles.

I may not be a Michael Palin though I have been fortunate to travel extensively in Europe and the

Mediterranean countries with the occasional trip to China, Russia, Canada, Africa, Italy, Norway and the Americas. I just love travel and to be able to see so many sights of this wonderful world, but now in my dotage, I intend to concentrate my travels to the UK and its coastline. I have a particular interest in churches, cathedrals and castles, and there are so many other treasures worthy of my attention as long as health permits and I am fit to drive.

Chapter Twenty Seven
Family and Friends

My story would not be complete without mentioning those dear friends who, in their turn, have meant so much to me and offered me so many kindnesses over past years. First there was Ted Byron who befriended me during my teenage years when my need was at its greatest, and continued to do so for over forty years.

Whilst in Ipswich back in 1952 Don Rose, a newsagent, was an Ipswich Speedway supporter at Foxhall Stadium and it was due to my interest in the sport that I met him. Together we recruited a nucleus of like minded folk and formed a Speedway Supporters Club. Selling programmes, badges, clip boards, pencils etc., we created a roaring trade and with crowds exceeding twenty five thousand, sales rocketed. Sales kiosks were installed at each corner of the race track and more volunteers joined us. By now Joyce and 'Jimmy' (Don's wife who never used her real name, Mona) had become great friends and for many years we went on holiday together. We also shared a small cabin cruiser which we sailed on the River Deben at Waldringfield, midway between Woodbridge and Felixstowe. Many happy weekends were spent cruising down the river on a Sunday afternoon. Sadly Don died shortly after we moved to Melton Mowbray.

In 1960 I met Ken Gore Graham, an ex naval officer and a local bank manager, who I first met in the lounge bar of the Bell Hotel in Melton Mowbray. He appeared to be somewhat of a 'loner' who strutted into the bar with a superior gait but was quite amiable when I first approached him. He lived alone and introduced me to drinking pink gin. On his retirement we spent a holiday together in Malta, both having experiences there during the War. We got on well and in subsequent years made a couple of trips to Yugoslavia.

Another great friend was Ivan Kovach, a technical director with British Enkalon, a Dutch/German company in the 'man-made' fibre industry. He travelled extensively and was only home at weekends but usually drank in the Bell Hotel on Friday nights. Often I would be invited to his home in Warwick Road to join him in a few scotches. Sadly being a workaholic and heavy drinker, he died prematurely at the age of fifty-six.

John Roper, a retired Inland Revenue officer and a University of London graduate though years older than me became a great travelling friend. Very knowledgeable in geology and archaeology, he fired my interest in these subjects. He too lived alone and was a Rotarian. He loved the countries of the Mediterranean in which we travelled on many occasions, sometimes accompanied by his daughter, Diana. He, like me, was a member of the National Trust and we visited many of their properties together during the summer months.

Bim Bellamy whom I had known for many years when he was a representative in the Agricultural Machinery trade lost his wife shortly after Joyce died. As he lived at Morton near Bourne in Lincolnshire only twenty miles from Melton Mowbray and being in similar circumstances we became great friends, and still remain so.

All my friends with the exception of Ted and John Roper were Freemasons and we met regularly at Lodge meetings, which brings me to Jim Conway. He was at Petfoods Limited in Melton and lived at Burton Lazars and for many years bred dalmations which he entered for Crufts dog show. Jim was a loyal friend and enthusiastic in all his masonic activities, and encouraged me to join various Christian Orders. A few years ago Jim moved to Grantham but continued to keep up all his Melton connections. Sadly he died two years ago but I do, of course, keep in touch with his widow.

Cyril Hack is in a category of his own for not only did I work with and for him for nearly forty years, he remains a very special friend. Although he now lives in Kent there is scarcely a week goes by without us chatting on the phone. I have an open invitation to spend a few days down there whenever I feel so inclined. He lives with Pat Clark, the authoress who writes under the name of Claire Lorimer. This book would never have materialised had not Pat inspired me to write it and done so much to act as the editor to my original text.

Finally I must bring the reader up to date with my family. Robin our eldest son is now fifty-five years old and with his wife Helen, lives at Whittle-le-Woods, near Chorley, Lancs. He is Managing Director of a textile importing Company and spends much of his time in China, India and Vietnam purchasing raw materials.

Robin and his first wife had a boy and a girl, Anthony and Jenny. Both went to university - Anthony to Swansea and Jenny to Belfast. After Anthony had celebrated his 21st birthday at Swansea, he disappeared and two days later was found drowned on the Mumbles beach. It was a terrible shock to us all and at the inquest, an open verdict was returned by the Coroner. Anthony was such a happy lad, was captain of the football team and leader of a church band. It happened ten years ago and his death has remained a mystery.

Jenny graduated at Belfast University and I was a very proud grandad who witnessed the ceremony. Later she married an Anglican minister who, at the time, was a curate in Ballymena. A year later he became priest-in-charge of a parish in Belfast. He then joined the RAF and was commissioned to be Chaplain. After a two year spell at RAF Wittering the couple are now at RAF Station, Stafford.

Christopher, our second son, now fifty one years old, is married to Irene and lives in Smethwick. He is a Supervisor with Parceline. They have a daughter, Tina, who lives with her partner Brendan and they have a little girl, Ella-May, now fifteen months old. That made me a great-grandad and so at eighty-two, is there any wonder that I begin to feel my age?

Our only daughter, Julie, born in 1962, was educated at the Ockbrook Moravian School for Girls. On leaving school she joined Barclays Bank and is now the Personnel Manager at the Melton Mowbray Branch.

She married Stephen Gamble, then a Shop Manager for Hepworths, at Market Harborough. They have twin boys, Michael and Thomas, now aged eight and Sarah who is three years old.

Chapter Twenty Eight
"Minerva" – maiden voyage

Having enjoyed several cruises with Swan Hellenic on the *Orpheus,* now being retired, I took the opportunity to book a reservation on the maiden voyage, in autumn 1999, of her replacement named *Minerva* - the Roman goddess of wisdom.

The ice-breaking hull was laid in Odessa, the Black Sea Ukranian port of the USSR and was later taken to Genoa, birthplace of Christopher Columbus. It was built as a luxury cruise ship to be named *Minerva.* Of twelve and a half thousand tons and with a range of five thousand miles, she had a maximum speed of 17.5 knots.

John Roper and I travelled together and flew in to Genoa from Gatwick, then drove to the awaiting ship. There, in the reception lounge a Champagne party was in progress. With bands playing and the whole area bedecked with flags and bunting fluttering, we were conducted to our respective cabins. Emerging a few minutes later I had my first experience of the traditional 'razzamatazz' that was laid on at the commencement of a maiden voyage. Scores of ships, large and small had fire hoses and sirens going at full blast, three local bands on the dockside played and hundreds of spectators assembled throwing miles of mult-coloured ticker-tape as we left our mooring.

With low cloud and drizzling rain we sailed south to Naples, our first port of call but, arriving there next morning, the weather had not improved and although a visit to the Vesuvius volcano was planned, it was abandoned.

On board we had learned that the night before departing Genoa, the ship had suffered a massive daring robbery during which furniture, televisions and all electrical equipment and cutlery were taken. Most items had been replaced which explained why so many electricians were frantically fitting 13amp plugs to toasters, coffee makers etc., the next morning. The only deficiency was soup spoons. This was not the only problem as the cabin and restaurant crew were Ukrainians, Poles or Russians, many who spoke very little English. However, after the initial period they settled in and problems were overcome.

Leaving Naples we sailed towards Capri when suddenly the ship came to a halt and an announcement was made over the public address system; 'Attention all crew. Close all water doors.' Perhaps it was an exercise, thought some of the passengers whilst others showed concern less something untoward had occurred. An hour later and with no further announcement our journey continued.

On arrival at the spot where smaller craft should have taken us ashore to Capri we were informed that as it was a Bank Holiday, the ferry service was cancelled. Again, very disappointed I began to wonder if a certain 'jinks' might not be with us.

We had no alternative but to proceed on our journey

but then experienced yet another problem. The deck crew were Philippinos and were unfamiliar with their duties. This was borne out by the winch-man who, when dropping the anchor, had allowed the chain to over-run it, resulting in the anchor resting on top of the chain. All sorts of difficulties ensued when the anchor had to be raised. Half an hour later we eventually got going. But not for long as again an announcement was made: 'Attention all crew. Close all watertight doors.' There was another hour's delay before the engines burst into life.

Later that day I encountered the Chief Engineer, whom I had met previously, and enquired the reasons for the unscheduled stoppages. He informed me that on both occasions a main fuel pipe had fractured and the engine room was flooding with diesel fuel, but the problem was now satisfactorily resolved.

We continued south in the Tyrrhenian Sea, through the Straits of Messina, the channel of the Mediterranean Sea between Sicily and mainland Italy. The Straits connect the Ionian and the Tyrrhenian seas and is twenty miles long. At Messina a fleet of coaches awaited us and drove the forty miles to Taormina through absolutely magnificent scenery. Coaches were not allowed in the town due to the narrow, winding streets, and walking up the steep hill with temperatures soaring beyond seventy degrees Fahrenheit, we found it really hard going.

We were taken to the Greco / Roman theatre and then to a museum of highly decorated carts and carriages which were paraded on certain festive occasions. After a spell for coffee and shopping, we returned to the coaches and drove back to *Minerva.*

After sailing through the night we arrived at Katakolo (Peloponnese / Greek Mainland). Then on to the coaches which took us to Olympia. This is not a town but an excavated site where, in the 4th century BC, the Olympic Stadium was built. The distance between the starting and finishing lines of the track was six hundred feet, and forty thousand spectators could be seated on the surrounding grassy banks. Building of the Temple of Zeus nearby was commenced in 417 BC, and completed in fifteen years. Its ruined state is attributed to an earthquake in the early 6th century AD.

In one of the inner rooms stood the Statue of Zeus by the sculpture Pheidias - one of the Seven Wonders of the Ancient World. It was about forty feet high (seven times life size). Zeus sat on a throne of ebony and ivory decorated with gold and jewels. In his right hand he held the symbolic sword of Victory and in his left a sceptre and eagle. The statue remained there for a thousand years until the 5th century AD when it was probably taken off to Constantinople and destroyed by fire. In more recent times a village was built which has a hotel and a colony of shops for the benefit of the tourists.

The streets were lined with jacaranda trees in full bloom providing shade and a most beautiful background to this isolated area. Returning to the ship, we had just cast off and were easing away from our mooring when a Mercedes taxi roared up the dockside and the passenger raced towards *Minerva* frantically waving his arms. I was standing on the promenade deck, just above where the Captain was controlling the ship's departure, when I heard him exclaim:

"He's one of my crew. We shall have to go back." Fortunately *Minerva* was fitted with thrusters which enable the ship to be manoeuvred sideways so that little time was lost, but ships captains hate having to return to the dock and without a doubt the offenders were in for a severe reprimand in due course.

Our next port of call was Gythion in the southern Peloponnese where the coaches took us on a mountainous route affording magnificent views to Sparta. The city-state dates back to the 11th century BC when seven year old boys were trained for war. We visited the Shrine of Artemis Orthia by the River Eurotas and the museum, and were shown face masks dating back to 635 BC and lots of carved ivory and bone objects. The following day our destination was Heraklion some one hundred and fifty two nautical miles away.

After breakfast John and I took the excursion to Knossos and visited the Palace of Minos where the story of the monstrous Minotaur was told. The site was excavated by Sir Arthur Evans from 1900 AD till his death in 1941. Many findings were dated by radiocarbon to 6000 BC. This tour was very exhausting as temperatures were over seventy degrees Fahrenheit and there was a lot of walking on uneven ground. We returned to *Minerva* for lunch but instead of completing the tour, I rested and watched the Wembley Cup Final between Liverpool and Manchester United on TV. Cantona scored the only goal of the match in the eighty-fourth minute and Manchester United won the Cup.

Another one hundred and sixty six nautical miles took us to the island of Rhodes. The sea was as calm as a mill pond and with only a light breeze, the journey was most enjoyable. Rhodes is the largest island in the Dodecanese and well known with its connection with the Knights of St John who stayed here for two hundred years. We visited the Hospital of the Knights and the Infirmary Ward with its seven octagonal pillars.

Walking down the Street of the Knights on our way to the Grand Masters palace we observed the various Coats of Arms sculptured over the doorways. The Knights Hospitallers having been driven from the Holy Land in 1291 AD, arrived in Rhodes, transformed the fortifications and withstood the Turkish Siege of 1522 AD which was led by Sultan Suleiman, the Magnificent who lost 90,000 of his 100,000 men. The Knights were then allowed to sail away to Malta.

When the tour ended I made my way down a side street and saw a plaque on a wall commemorating the execution of three hundred Jews by the Germans in World War II. The door nearby led into a Jewish synagogue and never having visited one hitherto, I entered and was shown round by the lady custodian. As I was about to leave, the custodian asked me if I would do her a favour. She needed to leave and do some shopping she told me. Would I take care of the building while she was away? I readily agreed and no sooner had she gone, than a party of Americans entered. I showed them round and repeated what the custodian had previously told me and they were impressed. I have since wondered if that was the first time that an Anglican had been left in charge of a Synagogue!

Returning to the ship before sailing I watched our departure and soon realised that the crew – more experienced now - were much more responsive to the orders given, and procedures were as they should be.

Our next destination was Kusadasi, some one hundred and thirty nautical miles from Rhodes, and we arrived at 7am. The excursion I had chosen was to encompass the whole long and gruelling day with three sites to visit and a lot of walking. I shared a seat in the coach with Richard Chartres, the Bishop of London and one of our lecturers, who was extremely good company. The outskirts of Kusadasi were a mass of modern tower blocks with holiday homes scattered on the hillside. Passing through spectacular mountainous country we reached Priene where we started our ascent to the site of the Council Chamber of Bouleuterion one thousand feet above the Meander Valley. There, cut into the hillside, were stone seats on all sides to accommodate up to seven hundred people.

In the centre stood an altar from which Richard Chartres described the area in great detail. Climbing higher we saw the remains of the Temple of Apollo and still further, the Temples of Athena and Zeus. Flora in this location was magnificent with the vivid contrast of dark blue irises and red poppies in profusion.

Miletus, situated at the head of the River Meander, was our next call. Once an important maritime trading city with four harbours, three for commercial traffic and one for pleasure, it faded into insignificance as the river silted up. The 3rd century theatre came next followed by the Baths, Ionic Stoa and the dilapidated Mosque of Ilyas Bey where wild flowers had overgrown the neglected graveyard. Driving on to Didyma we had lunch in a restaurant opposite the Temple of Apollo.

After an excellent meal we crossed the road on a conducted tour of the Temple, the largest in the world. Three hundred years were spent in its construction but it was never completed. Most was destroyed in subsequent earthquakes but much remains including three enormous columns, twice the height of those in the Parthenon. The Oracle was a separate building in the centre of the site. One could have spent hours here but time marched on and we had to return to Minerva.

Our next destination was Dikili, one hundred and nineteen nautical miles on. Arriving at 12 noon the following day, we were taken ashore on a lifeboat to join the coach to take us to Pergamum. After yet another hard climb we reached the Acropolis, over a thousand feet high. Then on to the Asklepieion, named after the God of Healing where the last of the pagans had left in early Christian times. Most remaining buildings were Roman and the columns Corinthian. Following destruction by earthquake they were replaced with those of the Composite Order. There was a small theatre to hold about three and a half thousand spectators, and a public latrine for about forty, similar to the one in Ephesus. A Pump Room with an intricate water system supplied stone baths used for water healing.

Returning to *Minerva* we sailed for Delos, a distance of one hundred and thirty six nautical miles. A tiny island comprising 1.3 square miles in the Cyclades Group, Delos was colonised in the 10th century by the Ionians who developed the cult of Apollo. The Temples of Apollo, the Temple of the Athenians and the Porinos Naos - the three principle temples - stood side by side in the centre of the precinct. The Avenue of Lions carved in Naxian marble dated back to the 7th century BC. Near to the theatre we saw a complex of well excavated and restored private houses laid out along winding streets.

Uphill were two houses of interest, the House of the Maska and the House of the Dolphins. The highest point in the island we were told, was Mount Kynthos, on which had been built a number of monasteries some still inhabited.

Back to 'Minerva' and on to the Port of Pirieus where the harbour was crowded with forty other ships and twenty others awaiting a berth. Two hours later we were safely moored and I was able to obtain a ticket for the Mycenae / Epidauros excursion. As we drove by coach out of the city in heavy traffic, we crossed the bridge over the Corinth Canal and stopped to take photographs. Half an hour later we reached Mycenae.

During the late Bronze Age (c.1600-1100 BC) Mycenae was the most important centre of the Aegean civilisation on the Greek mainland. Climbing a steep hill we entered the city by the Lion Gate. We continued uphill to see the Royal Shaft of Graves (16th century). These were circular with domed roofs and called 'tholos' but often referred to as beehive tombs and were used by royalty. Palaces and houses built on the mountain summit were farther ahead but as the road was cobbled and it was already raining, I returned to the coach.

After lunch in a roadside restaurant, we drove on to Epidauros. Climbing further uphill to the theatre, the local guide described the history of the area and one of our party read an extract from Shakespeare to demonstrate the excellent acoustics. We returned to the Port of Piraeus by the coast road from which we enjoyed magnificent views during the one and a half hour journey. We returned to *Minerva* and sailed at 7.30 pm for the Corinth Canal, some thirty-one miles away.

The sea was calm and it was a beautiful moonlit night; perfect conditions to negotiate the Canal. Apparently *Minerva* was the largest ship to negotiate this passage. While still a mile from the Canal entry, two pilots were taken aboard and a tug positioned forward of *Minerva's* bow. At midnight and under a full moon we proceeded at a very slow speed, estimated at 2 knots. With only a few inches to spare on either side, we crept forward. The canal is just four miles long and our passage took one and a half hours. Most passengers assembled on deck to witness this fantastic control of *Minerva* which was achieved under her own power and with no assistance from the attending tug. This was only possible in perfect weather conditions. We were lucky and I shall ever remember that exciting evening.

Minerva arrived at Itea at 6 am., and after breakfast we joined the coach bound for Delphi. For fifteen miles we climbed nine hundred feet along winding roads to Mount Parnassus. Delphi, sacred to the Greeks, was believed to be the 'omphalus' (navel or centre) of the earth and this was designated by a large conical stone. The Pythian Games were held here every four years.

We had already climbed nine hundred feet to the Temple of Apollo with the adjacent Delphic Oracle, and as rain threatened and we had seen the Temple of Athena on the way up, I returned to the coach where I found my travelling companion, John Roper fast asleep. There was a limit to the amount of uphill climbing with which we could cope. On our return to '*Minerva*' we had dinner and I retired at 10 pm.

Sunday morning I rose early and attended a Service in the Main Lounge conducted by Richard Chartres assisted by Colin Babcock of Winchester Cathedral. In the evening John and I were among those chosen to share the Captain's table for dinner.

Venice was reached on the final day and after a flying visit to St Mark's Square and the Doges Palace, sufficient time was left for shopping. We then travelled back by water-bus to the airport to catch our return flight to Heathrow, and so ended a most interesting cruise, but oh, so tiring!

Chapter Twenty Nine
The Silk Route and China

It was when Robin, our eldest son was living in Shanghai in 1995 that I decided to take the railway cruise from Russia to China which ended in Shanghai, and then spend a holiday with him. This was a daunting undertaking but John was keen to join me. In the event he was taken ill some weeks before we were due to leave and he was forced to cancel so I travelled alone.

Flight 236 was non-stop to Tashkent in Uzbekistan and was followed with an internal flight to Samarkand. Here our group stayed at the Samarkand Hotel and after freshening up and a meal, were taken on a conducted tour of the city.

This is a city on the River Zeravshan. It had many industries and a population of 366,000 (1989). A sight –seeing tour included the architectural monuments in Registan Square and the Guy-Emir Mausoleum. This was the 15th century Moslem style burial place of Tamerlane, the great warrior of the Middle Ages. Subsequently we were driven to the station to board the Gorbachov train, but it was not to be found. We toured around the extensive marshalling yards and eventually found the train half a mile away. With no platform in sight and the driver not willing to move it, there was no option but to trundle along the rail track carrying heavy luggage and scramble aboard. Fortunately, a much younger passenger seeing me struggling, relieved me of my heavy case and helped me aboard. The compartments were well arranged in pairs, one for sitting and sleeping and the other for washing and toilet, all very satisfactory. A female attendant occupied a compartment at the end of each coach and was always available if help was required.

Two enormous steam locomotives were hitched on and we were soon on our way. Our first stop was at Tashkent and after a sight seeing tour we rejoined the train to travel through the Kazakhstan Steppe, a vast expanse of unproductive land. Lunch was served in the restaurant cars. Alma-Ata (Father of Apples), the capital of this newly independent state and the seventh largest country in the world, has a population of 1,068,000 - 70% Russian and only 12% Kazakh. It was founded in 1854 AD., on the banks of the Almaatinka in the foothills of the northern Tien-Shan (Heavenly Mountain) range. The name of the city derived from the apples that were grown in the river valley.

We continued our journey northwards during the night and reached the border town of Druzhba at dawn. Then began the drawn out procedure of crossing the border into China. After breakfast we were informed that all restaurants and toilets would be closed until we had completed the transfer. Little did we know that these procedures would take over seven hours. First, two Russian officials collected all passports; an hour later a transit form was issued by two others. Two hours later the forms were collected and eventually our passports were returned. The train then moved into an area known as 'No Man's Land'. There it stayed for some time. All passengers were then instructed to assemble on the Russian side platform. A white line marked the extent of Russian territory and woe betide anyone who stepped over it! After an interminable wait, instructions were issued for us to cross over to the Chinese platform, similarly white lined, before we eventually boarded the Mao Tse Tung train. This was more luxuriously furnished and was spotlessly clean.

We were served excellent food in the restaurant cars which were painted red, and separate carriages painted blue provided numerous shower facilities and were available twenty four hours a day. A uniformed official attended each coach and on arrival at a station, would don his cap, open the carriage door and stand to attention on the platform until the signal was given for the train to move away. The railroad was quite fantastic as it was tunnelled through the mountains and bridged across the rivers and ravines. In many respects it reminded me of the Rocky Mountaineer train in Canada. However as far as we were concerned there were a few unscheduled stops. From time to time we were shunted into a siding. There we stayed until a freight train passed by causing a delay of anything up to a couple of hours. Apparently these commercial trains carrying coal, oil or steel always took priority.

We were now on the Silk Route, and at Jiayuguan, made a short stop to see the western end of the Great Wall of China. This, the western extremity of the Wall, stretched almost 6,000 kilometres to the Shanghai pass in the east. Three separate walls built in the former states of Qin, Zhao, and Yan were connected and reinforced in the Ming Dynasty (1368-1644)

On the second day of our rail trek we approached the Gobi Desert, (500,000 square miles). Four thousand feet above sea level it consisted chiefly of sand and gravel plains, broken by low rocky mountain ranges and salt pans. We were informed that the fringe of sparse pasture land was inhabited only by Mongolian nomads.

Passing through the Gobi Desert via the city of Urumqi, we reached the junction for Turfan. Here we left the train and boarded coaches to continue the journey to the oasis city of twenty thousand people. As Turfan lay at the foot of the Flaming Mountains there was no water shortage, and grape vineyards and apple orchards thrived.

Many centuries ago nomadic tribes dug for water in the area now known as Karez. They had observed the melting snow on the distant mountains and deduced that some, at least, must flow under the desert. It was not until they were a hundred feet down that water was found. Many such wells were made and a tunnel dug downhill to where the town of Turfan (154 metres below sea level) was built. This together with fertile soil allowed vineyards and orchards to be developed in what had hitherto been a barren desert.

Trees lined the streets as we drove to the Turfan Hotel where we stayed overnight. We toured the city and passed down 'Grape Valley' before returning to the train to continue our journey towards Liuyuan.

As we completed our crossing of the Gobi desert, the

terrain changed to fertile soil and together with the more temperate climate, allowed crops to be grown. Land was divided into small holdings of either one or two hectares. All work was done by hand with an occasional ass or mule here and there, and both man and wife laboured for what must have been a pittance.

Just thirty-one miles on was the ruined city of Gaochang which, in its heyday had thirty thousand inhabitants and thirty Buddhist monasteries. Our next stop was Liuyang, home of the Naxis people. Here the custom was for the women to choose their partners in festive rituals. The chosen man only spent the nights with 'his' woman for many years; during the day he continued to work and live in his mother's house. Female members of the family had the privilege of their own room where they could receive their lovers, while the men had to make do with shared rooms.

Crossing the Yellow River at Lanzhou and skirting the edge of deserts, we reached the oasis town of Dunhuang where cotton was produced. We left the train there and were coached fifteen miles to the Mogao Caves, the first of which dated back to 366 AD., and the last ones carved at the time of the Mongolian Conquest in 1277 AD. Over forty thousand manuscripts alone were found in one cave, and in the grottos, an uninterrupted history of landscape paintings over a period of a thousand years.

In the evening an extra excursion took us to the site of the Crescent Lake, a site that was claimed to be of exceptional beauty. Barriers had been erected about a mile from the lake and we had the choice of hiking, riding on a camel or in a camel cart. Not feeling inclined to walk or suffer the uncomfortable camel ride, I opted to join a queue for travel in a camel cart. Only two carts were operating each taking four passengers which resulted in a long wait. At last I was at the head of the queue followed by two middle-aged American ladies who took the first two seats. I sat in one of the rear seats out of harm's way recognising the proximity of the front seats to the camel's posterior. Not only was the ride a most uncomfortable one but the front seat passengers had the benefit of the exhausting sound and fumes emitted by the aged camel.

On arrival at Crescent Lake it was found to be empty so the whole evening was a waste of time. I walked back to the coach under a setting sun and was glad to be aboard the train once more.

Our journey continued on to Xian, a modern industrial town with three million inhabitants. We toured the town passing the Great Wild Goose Pagoda, the Drum Tower and the Great Mosque on our way to the newly discovered and important sight of the Army of Terracotta Warriors. These dated back to the first Chinese emperor, Qin Shi Huangdi.

In 1974 peasants digging a well, uncovered life size horses and almost seven thousand Warrior figures. These had since been restored and were now exhibited in a hall built above the excavation site. The figures were arranged in battle formation in eleven columns - soldiers holding spears and swords, others steering horse-drawn chariots: and there were officers, all figures about five feet six inches tall. Each head had been specially modelled with individual facial expressions.

Other items on display included swords, dagger-axes, spears, halberds, hooks, battle-axes, long spears, bows and arrowheads.

In a separate hall we were shown a miniature model of a bronze chariot with horses and coachman from the Qin Dynasty. One could only marvel at these fantastic sites, the only disappointment being that our visit was limited to one hour and the use of cameras was forbidden.

On leaving the exhibition we transferred to the excellent Golden Flower Hotel and after settling in, had lunch before departing on a sightseeing tour of Xian, China's earliest capital. We visited the Provincial Museum displaying objects from pre-historic times as well as recent finds unearthed from ancient tombs. We included the City Wall and the Big Wild Goose Pagoda, erected in the 7th century, before returning to the hotel.

In the evening we were taken to an exclusive restaurant for a 'Dumpling Dinner'. There we sat at circular tables fitted with a glass turntable accommodating ten diners. A compere announced there were to be twenty-four courses, each one comprising a small dumpling with a different filling. As the waiter served ten dumplings on the turntable, the compere described its contents. They included various fish, meat or snake portions and we chose the ones that tickled our fancy. Most diners sampled probably ten dumplings but I decided to be very selective and was satisfied with seven which I thoroughly enjoyed. It was an interesting evening and apparently unique to Xian.

Continuing our train journey next day, we arrived at Luoang and toured the Longman Grottoes, a colossal example of religious art. Construction of the two thousand one hundred grotto niches, forty three pagodas, three thousand six hundred tablets and steles and more than one hundred thousand statues lasted over four hundred years. The smallest statue was two centimetres tall and the largest seventeen point four metres. It was a most impressive visit although we only saw a fraction of the grottoes.

Crossing the Yangtse River Bridge later that day, we reached China's oldest southern capital of Nanjing surrounded by the Purple Mountains. This bridge was opened in 1968, and took nine thousand workers and eight years to build. It was five thousand one hundred and eight feet long and two-tiered with car traffic on the upper level and rail traffic below.

Although snow sometimes fell there in winter, the temperature rose to over forty degrees centigrade in mid-summer. With only a short stop there, we saw very little of the city. Our journey took us on to Wuxi passing many fields of mulberry bushes. For almost fifteen hundred years, Wuxi was the centre of silk production. Most peasants cultivated silkworms as a popular and profitable sideline. The young worms, spread out on rice straw mats, were fed with juicy mulberry leaves and then placed on bundles of rice straw where they spun a cocoon within five days. The cocoon was then washed in silk spinning mills in a hot water flume. At one of these factories we watched girls scooping the cocoons out of the water with their bare hands working twelve or more hours a day for wages equal to four pounds a month. After washing the silk, thread was pulled, each cocoon producing three thousand two hundred and eighty feet or more of thread. Wuxi enjoyed a mild climate and with sufficient water and fertile soil, it is one of the most productive regions in China.

We visited the Xihui Park encompassing one hundred and eleven acres in the western part of the town at the foot of Tin Mountain. Then on to the Taihu Lake into which flowed the water from the Hui Mountain.

From Wuxi we travelled by boat for the next six hours along the Emperor Canal to Suzhou. The canal, which connects the rivers Haihe, Huanghe, Huaihe and Yangzi was forty miles long enabling trade to be developed between north and south China. Rice was transported from the south and as a result, the fertile regions of southern China gradually developed into an agricultural centre. This was an extremely busy waterway with hundreds of commercial barges travelling in each direction as well as the expanding tourist trade.

Under a cloudless sky and with a gentle breeze we sailed the length of the canal and four hours later arrived at Suzhou known as the town of gardens and canals. The most prosperous period was in the Ming and Qing Dynasties (6th century onwards) when officials, scholars and artists settled there and local traders grew rich. This wealth was largely invested in the one hundred and fifty gardens which made Suzhou so famous.

The principle of Chinese garden construction is creating an illusion of the universe in a small space which can be clearly seen in these gardens with water flowing between bizarre, rocky shores, connected by canals and zigzag bridges, winding paths and craggy rock formations. We explored the Garden of the Foolish Politician which covered nine acres and is the largest garden in Suzhou. Although it was drizzling with rain, I enjoyed seeing ponds filled with lotus flowers and the willow tree lined pathways.

Finally we returned to the Suzhou railway station, boarded a train and in less than an hour arrived in Shanghai. The station there was huge but rules were very strict. All arriving passengers were obliged to leave the platform immediately and assemble in the open area adjacent to the car and rickshaw parks. Luck was not with me for although my son, Robin met me as arranged, it was in torrential rain that I stood getting drenched to the skin whilst the hire of a rickshaw for the luggage was negotiated. Next time, I told myself I must remember to have a folding 'brolly' in my hand luggage. We were driven to our hotel in a mini-coach but the traffic in the city centre was horrendous.

In Shanghai we stayed at the JC Mandarin Hotel and I lost no time in unpacking, having a bath and changing into dry clothes. Our arrival coincided with the onset of the monsoon season and we experienced torrential rain far heavier than ever seen back home. There was a short respite after dinner and we took the opportunity to find our way to the Bund, famous throughout China for its many shops.

Shanghai is the largest city in China having a population in excess of thirteen million, and seemed to me to be the most congested in the world. Built on the side of the River Hyangpu, the name Shanghai means 'up river to the sea.' Traffic is a mixture of buses, cars, taxis, rickshaws and thousands of cycles many with a side-car arrangement laden with twenty feet lengths of metal or timber. Our journey from the airport to the city centre, approximately ten miles, took almost two hours in mid-morning.

We planned to spend only two days here, sufficient time to get a general impression of the modern and ancient parts of the city and to visit the enormous street market. An interesting daily event occurred each morning when, at eight am., loudspeakers broadcast music and all the pedestrians on their way to work adjourned to the nearest open space and engaged in aerobic exercises which included 'shadow boxing'. At precisely eight thirty the music stopped, exercises ended and all went about their business.

We left Shanghai bound for Hangzhou on a train with double tiered coaches at seven am in pouring rain.

Seats were booked on the upper deck affording us a wonderful view of the surrounding agricultural area where the crops comprised one of wheat and two of rice each year. Three hours later we arrived in Hangzhou and were met by some of Robin's friends. Robin has traded in the Chinese textile industry for many years and most of his suppliers had their factories in this area.

All the major trading companies in China have a separate Hospitality Department whose sole duty is to meet customers on arrival and provide transport during their stay, book hotels and organise entertainment, including sight-seeing trips, and arrange their business meetings. A driver with his vehicle and a courier was allocated to each customer as they arrived. It mattered not how many were in the visiting group as a suitable vehicle was always found. The day ended with dinner at a five star hotel or restaurant followed by karaoke and drinks after which we were taken back to our hotel and times were arranged for our collection the next morning.

Hangzou, the capital of Zhejiang Province is one of the most prosperous regions of China, famous for its silk production and Dragon Well tea. Its population is 1.2 million. Tourists were attracted to this city particularly to West Lake, which covered an area of two square miles. Here we were met by representatives of one of Robin's major suppliers.

After a tour of the city we were driven to West Lake to the north-west of Hangzhou. Of the thirty lakes in

China this was the largest. There were three islands within the lake, the largest of which was known as Solitary Hill Island. Specially constructed pleasure boats with dragon's heads on the bow and stern and carrying about thirty passengers plied for the tourist trade. A full day tour had been arranged for us which allowed sufficient time to see each of the islands and to have lunch on Solitary Hill Island.

First we walked along the Long Corridor, a small light wooden structure decorated with countless painted scenes from Chinese mythology, which ran parallel with the northern shore of the lake. The Su Causeway with its seventeen arches led from the northern shore, near Hangzhou Hotel, to Flower Bay Park in the south-west. That spring day the Park contained many rare flowers, peonies, rocky landscaped areas, pavilions and numerous small fishponds. Another unusual sight was the Marble Boat, a full size replica of those on the lake. With glorious weather under a cloudless sky we thoroughly enjoyed this trip, away from the madding crowds and with such beautiful scenery.

On leaving the lake we were on our way back to the Golden Summer Hotel when the courier enquired if we might be interested in seeing an Ice Exhibition which was being staged only five miles away. We eagerly agreed and duly arrived at the site where huge marquees, equivalent in size to four circus big tops stood. The first section was an air lock area where we donned heavy Chinese army greatcoats, winter boots, hats and gloves and were warned that in the exhibition hall, the temperature was maintained at twenty degrees below freezing and that the tour would be limited to twenty minutes. On entering the hall the scenery was spectacular. A complete miniature village had been constructed in ice with varying coloured neon lighting embedded in each structure. The village was complete with roads, bridges, houses, temples and pagodas and was absolutely breathtaking. Apparently this was an ancient custom practised in northern China, and the first time it had been exhibited in this region.

The following day we proceeded to Lanzou, the capital of the Gansu Province, and stretching alongside the banks of the Yellow River. With a population of over two million this was a modern industrial city and the centre of Chinese atomic research. Here we visited a six hundred year-old temple situated on the slopes of the Park of Five Springs and the Chongqing Temple, and continued on to the Binglingsi Caves to view the huge carvings of Buddha in the sandstone rock face and dating back to the fifth century.

To my surprise we witnessed a modern image of a reclining Buddha being constructed - not of stone but of sheet metal. The front section must have been at least fifty feet long and twenty feet high. After being beaten into the desired shape and ready to be bolted to the rock face, it was sprayed with paint to which stone chippings were added to give the impression of a genuine stone Buddha. It was then heaved into place and fixed into position. One wondered how many of the others were genuine!

From Lanzhou we travelled on to Beijing (previously

known as Peking) the capital of the People's Republic of China, which had a population of over six million. The centre of the city was dominated by Tianan-men Square, named the Square of Heavenly Peace, said to cover five hundred acres. The largest city square in the world, it was where the notorious massacre of the students in the uprising of 1989 took place.

There we visited the Mao Mausoleum, the monument of the Heroes of the Nation, then on to the Forbidden City with its Imperial palaces, and the Hall of Supreme Harmony with the exquisitely carved and gilded Dragon Throne, once only accessible to the Imperial family and their staff. It was open to the public on payment of a heavy entrance fee. Although it was an uphill climb through the various palaces, temples and halls, these were sights not to be missed.

The toughest climb of all was the approach to the eastern end of the Great Wall of China, said to be the only work by human hands which is visible to the naked eye from the surface of the moon. This enormous structure, parts of which dated back to the fifth century, wound its way through five provinces. I had several glimpses of the wall from the train between the Gobi Desert and Shanghai and now I was at its eastern terminal known as Badaling. From here there is a breathtaking view of the surrounding scenery where the mighty wall climbs up and down, winding its way through the mountainous landscape.

This was certainly one of the highlights of my whole tour. The time had now come to prepare for my homeward flight but not before a visit to Hong Kong, the British Crown Colony where Chris Patten was negotiating the terms of the handover to the Chinese Nation (terminating the ninety-nine year lease agreed in 1898 with the British). It was to be designated 'A Special Administration Region of the People's Republic of China'.

We flew from Beijing, landing in Hong Kong on a runway in the centre of the city with the river on the left and skyscrapers on the right, both of which appeared to be but a few feet away from us. As we arrived in Hong Kong at the onset of the rainy season, we should not have been surprised to encounter torrential rain but, on this occasion, we had our 'brollies' handy. We checked in at the Holiday Inn-Harbour Lights Hotel in Causeway Bay and took a room on the sixteenth floor which overlooked the harbour. From there we had a bird's eye view of the junks, yachts and freighters entering and leaving, and the Star ferries crossing over to Kowloon. The ferries left at twenty-minute intervals and seven minutes later had completed the crossing.

On our first morning in Hong Kong the clouds hung low over the harbour and heavy rain continued. We chose to adjourn to the lounge and watch the activities in the harbour below where we noticed an area which had been roped off so as to be kept clear of all other traffic.

The barman told us this was to be the scene of the International Championships of the Dragon Boat

Races that would commence at ten o'clock each morning and continue throughout the week. Having never seen this sport before, we had much to learn. The boats used were long and narrow with a dragon figurehead, and carried a crew of twenty, each member being provided with a short-handled paddle which he wielded at the incredible speed of eighty to ninety strokes a minute. The competition was run on a knock-out basis and six teams competed over a distance of half a mile in each heat. Successful heats were run at fifteen - minute intervals each day until four o'clock. We found it to be a most ideal and exciting spectator sport which we observed on a wet day from the comfort of our hotel lounge.

Fortunately the weather improved after lunch and we were able to make our first exploits in to the city centre. We strolled round the Causeway Bay area and Nathan Road said to be the Mecca of rare antiques and designer labels, but made no purchases. We went on to Stanley Market - the Petticoat Lane of Hong Kong. This was a real tourist attraction as well as a source of supply for buyers from the eastern European countries and Russia. Here a visitor could haggle to his heart's content. I did a deal for four silk shirts which started at one pound fifty pence each and were bought at fifty pence after haggling for twenty minutes and walking away several times. It reminded me of trading with the Arabs in Cairo many years ago.

Our next visit was to the Jade Market where, trading in a similar manner, I bought several items of jade as presents to give to my family. Passing by the Aberdeen

Houseboat Community, we saw the local system of sun-drying of fish. We finally visited the Victoria Peak. Access is by means of the steepest tramway in the world and on reaching the summit one had a spectacular view over Hong Kong. Here we found extensive gardens, shops and a hotel, a real haven for every tourist. In brilliant sunshine we sat in the gardens having liquid refreshment without a care in the world. Alas! This was my last day of a memorable month in a part of the world unknown to me and an experience that I shall never forget.

We returned to the hotel, I collected my luggage and we made our way to the airport. I bade farewell to Robin, thanking him profusely for making this such an exciting and memorable time, and boarded the Virgin Flight back to London Heathrow.

Chapter Thirty
Review

To end my story I cannot resist the temptation to reflect on the events of past years. I was fortunate to be taught at an early age discipline and the difference between right and wrong. Like any other child I transgressed on more than one occasion and was severely punished. I felt the lashings of Dad's belt as a result of my misdeeds but I soon learned the intended lesson.

Another major factor in my formative years was the influence of those kind folks who befriended me. Lastly I was favoured with ambition backed by a strong belief in the Christian way of life.   These together with a lot of luck and the opportunities that popped up from time to time have made me what I am despite my limited schooling which finished at the age of fourteen. My real education has been learned in the 'University of Life'.

Having lost my wife Joyce sixteen years ago, I still live alone and without any outside help or care for the house and garden. Fortunately Julie and Stephen live only half a mile away and are always on call should I need help. Most Saturdays, and during school holidays I spend with the grandchildren and enjoy making simple toys or books or reading with them. On most Sundays I join them for lunch.

As I freely admit, I hate loneliness, so I concentrate on keeping busy either at home, at social activities, with my Masonic duties or visiting the aged, sick or bereaved - work which I find most rewarding. My health is good and I live a happy and contented life.

What more can I ask?

1. Geoffrey 2. Stephen 3. Tina 4. Brendan 5. Irene 6. Christine 7. Helen 8. Pat 9. Barbara 10. Donald 11. Jenny 12. Michael 13. Julie 14. Sarah 15. Thomas 16. Christopher 17. George 18. Robin

George's eightieth birthday party with family

Passing Thoughts

Life is made up of joy and also of sorrow,
Happy today perhaps sorrow tomorrow,
Some days it rains whilst others it snows,
I never know why for only God knows.

Some folk are joyful while others are sad,
Some make you happy and for that I am glad,
Some not so friendly and don't want to know,
That you may be lonely, know not where to go

But then the sun shines; the clouds seem to clear,
Along comes a friend and soon he is near,
The kettle goes on, let's have some tea
A cheerful chat that follows twixt you and me

Just like a patchwork life seems to be
Of various colours earth, land and sea,
Of people you meet and some travel far
To contribute much to make you what you are.

G. Hows
June 2004

Ode to our "Mungy King".

His name it is Johnson, a Corporal bold,
The feeding of `erks` is the power he holds,
He is "King of the Mungy", as tough as can be
Just watch him punch cards with Satanic glee.

Sometimes there`s grub over, we call it gash
Usually comprising some ghastly hash
But will this concoction to us be lobbed out?
No not whilst dear Mr Johnson`s about.

Sometimes for supper thick porridge they serve
And if there`s any left over and you`ve got the nerve
You can stand in a queue for an hour or so
Watching the dam stuff gradually go,

But should our dear Corporal be on duty that night,
He will glower and scowl and fill you with fright
"No gash tonight" will be his vile cry
"So back to your billet and lay down to die.

When this bloody war`s over won`t we be glad
Just picture our Corporal's son saying "Say dad,
What part did you play in that conflict so hard?"
"I punched millions of holes in dirty wee cards".

Geoff Smith 1435 Sqdn. 1942.

To "O"

To the memory of our dear "kite"
The veteran of many and many a flight
And to those who will never be in the know
Her letter was painted with a nice white "O".

Her symmetrical lines were a pleasure to see,
Her body and wings as clean as could be,
But the high spot of her beauty, the story runs
Lay in the power of her 'dirty' great guns

No words of mine can describe her to you
She was always just "pukka" through and through,
And there's tears in my eyes as these lines I write
For I know she'll no longer enter the fight.

She's "spitchered," finished, as dead as can be,
And never again will her beauty I see
She's "had it" and never again will she fly
So "Goodbye you old bastard Goodbye, Goodbye"

Geoff Smith 1435 Sqdn. 1942.

Ode to Spitfire Maintenance Flight.

In Spitfire Maintenance 'tis wizard they say,
Yes, that's what they told me the very first day,
That I arrived in Malta G.C.
Not knowing what was meant for me.

The very first morning we started at eight,
Inevitably, some of the fitters were late,
Then tool kits were drawn, to the 'pens' we dispersed,
But before very long did something we'd never
rehearsed.

"The wanker's just gone" said a fitter with glee,
"So out with your fingers and just follow me."
"Two six to the quarry, yes Safi I mean."
So we doubled across as though we were keen.

"The Camp wanker's gone, M.E's overhead"
Cried Flight Sergeant Thornet, as he had just read
In yesterdays paper the total score
Of kites destroyed, 'twas a dozen or more.

A single raider came in, dropped a couple of 'eggs'
Then out came the 'swaddles' as fast as their legs
could carry them carry them out to a sizeable spot
In the runway, where a 109 had dropped his lot.

The "All Clear's gone, the red flag`s down,"
Cried a fitter called Thompson with a hell of a frown,
"Back to the Spit in 124 pen
Get some more work done and learn some more 'gen'."

Of course, chaps on the Flight think we're a lot of
dim blokes
Though actually, of course, 'tis usually the joke
is on them, but we don't rub it in
We take all their 'ragging' with a jolly old grin.

They sit in their 'shacks' from pre-dawn to dusk,
We start after daylight and then we all must
be finished by eight and ready to go
to dear old Vallett, canteen and Lotto.

Their life must be a monotonous one
To sit on their 'backsides till set of sun,
With nothing to do but a 'mingy' D.I.
Or make nasty comments to Maintenance nearby

All the real work on the Spits, of course, we do,
We have to inspect them and clean them too,
For sometimes they come in a hell of a mess
And what's under those cowlings you'll never guess.

But we are the 'gen men', we know our stuff,
Believe it or not, but that ain't no 'guff',
A carb change we do and set it just right,
And that Spit's like a humming bird when next in flight.

No 'mag drop' or 'boost surge' when we've done the job
Or Chiefy or Eng-off would be ready to 'lob'
us 'over the wall' for a couple years
Without hesitation and still less any tears.

Yes we are the cream of the fitters, we know
And the blokes on the flights have a long way to go
But they are the boys who the 'kites' prepare
For another 'dog fight' with ME's in the air.

Dam fine set of chaps who are ever alert
For a 'Scramble' signal, no not the 'skirt',
First 'on the line' maybe second or third
And every man's kite takes off like a bird.

Yes, we're all in the 'mob', we all do our bit
To keep the kites flying and give Gerry the fit
If he comes over here any hour of the day,
We'll 'follow him down' or chase him away.

G. Hows. Luqa, Malta 1943.

A Reverie.

Sometimes when evening shadows fall
And the sun's last beauty passes,
I ponder then just to recall
How man a stupid ass is

He builds his house and plans his home
Prepares a perfect garden
From this he says I'll never roam
My heart to pleasure harden

Then comes one day a pretty lass
Whose form and beauty pleases
Does this same man just let her pass
While her attraction teases?

No, no not him the simple fool,
The sex in him is stronger
And he becomes a woman's tool
And is a man no longer.

G. Hows - 1435 Sqdn. 1943.

Notable Dates

Born	29th November	1921
Started School	January	1926
Apprenticed	January	1936
Enlisted in RAF	January	1941
Posted to Middle East	April	1942
Posted to Malta	May	1942
Demobilised	May	1946
Married	May	1947
David George born	February	1948
David George died	March	1948
Robin David born	June	1949
Christopher George born	September	1953
Joined Tractamotors as General Manager	April	1959
Julie Joy born	March	1962
Promoted to Sales Director	August	1976
Promoted to Managing Director	August	1979
Promoted to Chairman / Managing Director	December	1979
Retired as Chairman /M.D. but retained as Director	June	1987
My wife died	November	1988
Joined Melton Volunteers	January	1989
Peru Tour	March	1992
Euro Tour	April	1995
Minerva maiden voyage	September	1999
Elle May born to Tina & Brendan my first great-grandchild	December	2002

Masonic History

The Rutland Lodge No. 1130

Initiated	06. 02. 1964
Worshipful Master	01. 04. 1977
Past Provincial A.D.C	25. 11. 1983
Past Provincial Senior Grand Deacon	30. 11. 1990
Past Provincial Junior Warden	27. 11 1998

Vale of Catmose Chapter No. 1265

Exalted	17. 03. 1966
First Principal	18. 11. 1981
Past Provincial A.D.C.	25. 11. 1984
Past Provincial Sojourner	23. 03. 1990
Past Provincial Registrar	28. 03. 1996
Past Prov. Scribe Nehemiah	28. 03. 2002

Rutland Lodge of Mark Master Masons No. 1051

Advanced	11.10. 1967
Worshipful Master	09. 05. 1982
Provincial Standard Bearer	14. 05. 1985
Past Prov. Junior Warden	08. 05. 1990

Rutland Lodge of Royal Ark Mariners No. 1051

Elevated	24. 10. 1969
Worshipful Commander	10. 05. 1982
Provincial Grand Rank	12. 05. 1992
R.A.M. Grand Rank	10. 12. 2002

Oudh Chapter Rose Croix No. 241

Perfected	14. 01. 1974
Most Wise Sovereign	06. 04. 1986
30th Degree	09. 02. 1988
31st Degree	18. 05. 2004

Sir John Babington Preceptory No. 410

Installed Knight Templar	04 01. 1977
" Knight of Malta	01.11. 1977
Eminent Preceptor	04. 03.1986
Provincial Standard Bearer V.B.	10. 02.1986
Provincial Herald	12. 02.1990
Provincial Chancellor	10. 02.1997

Isle of Patmose Conclave No. 277

Exalted	26. 05. 1978
Appendant Orders	09. 04. 1979
Sovereign	04. 04. 1992
Standard Bearer (Constantine)	27. 09. 1985
Provincial 2nd Herald	12. 05. 1994
Past Divisional Registrar	13. 05. 1999
Divisional Senior General	11. 05. 2000
Past Grand Herald	03. 07. 2001

Order of the Secret Monitor. Amity Conclave No. 51

Inducted	17. 11. 1983
Supreme Ruler	25. 04. 1997
Provincial Bow Bearer	06. 06. 1998
Provincial Grand Visitor	02. 06. 2001
Provincial Guide	19. 06.2002
Past Grand Standard Bearer	13. 11. 2003

Leicestershire & Rutland Tabernacle No. 122

Admitted   27. 09. 1991

The Framland Hundred Lodge No. 9453

Founder member   27. 05. 1992

Progress Lodge of Installed Mark Masters No. 1786

Founder member   08. 11. 1997

Leics. & Rutland Chapter of Installed first Principals

Member   15. 03. 1996

Index

C

D

Z